# Review of
# Organic Functional Groups

*Introduction to Medicinal Organic Chemistry*

# Review of Organic Functional Groups

## Introduction to Medicinal Organic Chemistry

THOMAS L. LEMKE
*University of Houston*
*College of Pharmacy*
*Houston, Texas*

Second Edition

LEA & FEBIGER  •  Philadelphia

Lea & Febiger
600 Washington Square
Philadelphia, PA 19106
U.S.A.
(215) 922-1330

**Library of Congress Cataloging-in-Publication Data**


Lemke, Thomas L.
    Review of organic functional groups.


    Includes index.
    1. Chemistry, Pharmaceutical.   2. Chemistry, Organic.
I. Title.   [DNLM:   1. Chemistry, Organic.
2. Chemistry, Pharmaceutical.   QV 744 L554r]
RS403.L397      1988              615'.3              87-22810
ISBN 0-8121-1128-1

PRINTED IN THE UNITED STATES OF AMERICA


Print No.   3 2

# Acknowledgments

I would like to thank my colleagues at the University of Houston, Drs. Louis Williams and Lindley Cates, and at the University of Arkansas, Drs. Danny Lattin and John Sorenson, for their helpful discussion. I also wish to thank the SmithKline Corporation for its financial assistance through the Grant Awards for Pharmacy Schools.

In preparing the second edition of this book, I had the help of many individuals, especially students, who read the book and made suggestions on how the material might be better presented, and I would like to thank them also.

# Contents

*Contents*

# Introduction

This book has been prepared with the intent that it may be used as a self-paced review of organic functional groups. Were the material covered in this book to be presented in a conventional classroom setting, it would require 14 to 16 formal lecture hours. With this in mind, you should not attempt to cover all of the material in one sitting. A slow, leisurely pace will greatly increase your comprehension and decrease the number of return reviews of the material. You should stop and review any section that you do not completely understand. At the end of each section, questions are asked that you should be able to answer. If you do not understand the answer to any question, return to the appropriate section of the book and review it again.

## OBJECTIVES

The following outline is a general review of the functional groups common to organic chemistry. It is the objective of this book to review the general topics of nomenclature, physical properties (with specific emphasis placed on water and lipid solubility), chemical properties (the stability or lack of stability of a functional group to normal environmental conditions, referred to as in vitro stability), and metabolism (the stability or lack of stability of a functional group in the body, referred to as in vivo stability). There will be no attempt to cover synthesis, nor will great emphasis be placed on chemical reactions except when they relate to the physical or chemical stability and mechanistic action of drugs. This review is meant to provide background material for the formal pharmacy courses in medicinal chemistry. The objectives are presented in the following manner to aid in focusing attention on the expected learning outcomes.

Upon successful completion of the book, the following goals will have been attained:

  a. The student will be able to draw a chemical structure given a common or official chemical name.

  b. The student will be able to predict the solubility of a chemical in:
    1. an aqueous acid solution
    2. water
    3. an aqueous base solution

  c. The student will be able to predict and show, with chemical structures, the chemical instabilities of each organic functional group under conditions appropriate to a substance "sitting on the shelf," by which is meant conditions such as air, light, aqueous acid or base, and heat.

  d. The student will be able to predict and show, with chemical structures, the metabolism of each organic functional group.

To help you master these skills, the information is presented in the following outline form:

  A. **Nomenclature**
    1. Common
    2. Official (IUPAC)

  B. **Physical-Chemical Properties**
    1. Physical properties—related to water and lipid solubility
    2. Chemical properties in vitro—stability or reactivity of functional groups "on the shelf"

  C. **Metabolism—in vivo**
    Stability or reactivity of functional groups "in the body"

## RECOMMENDED PREPARATION

In order to maximize learning and to provide perspective in the study of the book, it would be helpful to read certain background material. It is highly recommended that a textbook on general organic chemistry be reviewed and consulted as a reference book while using this book. Pay special attention to the sections on nomenclature and physical-chemical properties.

# 1

# Water Solubility and Chemical Bonding

At the outset, several definitions relating to organic compounds need to be discussed.

For our purposes, we will assume that an organic molecule will dissolve either in water or in a nonaqueous lipid solvent. That is, the organic molecule will not remain undissolved at the interphase of water and a lipid solvent. If a molecule dissolves fully or partially in water, it is said to be hydrophilic or to have hydrophilic character. The word hydrophilic is derived from "hydro," referring to water, and "philic," meaning loving or attracting. A substance that is hydrophilic may also be referred to, in a negative sense, as lipophobic. "Phobic" means fearing or hating, and thus lipophobic means lipid hating, which therefore suggests that the chemical is water loving.

If an organic molecule dissolves fully or partially in a nonaqueous or lipid solvent, the molecule is said to be lipophilic or to have lipophilic character. The term lipophilic or lipid loving is synonymous with hydrophobic or water hating, and these terms may be used interchangeably.

Hydrophilic ........................ water loving
Lipophobic ........................ lipid hating
Lipophilic ........................ lipid loving
Hydrophobic ..................... water hating

In order to predict whether a chemical will dissolve in water or a lipid solvent, it must be determined whether the molecule and its functional groups can bond to water or the lipid solvent molecules. THIS IS THE KEY TO SOLUBILITY. If a molecule, through its functional groups, can bond to water, it will show some degree of water solubility. If, on the other hand, a molecule cannot bond to water, but instead bonds to the molecules of a lipid solvent, it will be water insoluble or lipid soluble. Our goal is therefore *to determine to what extent a molecule can or cannot bond to water.* To do this, we must define the types of intermolecular bonding that can occur between molecules.

What are the types of intermolecular bonds?

### 1. Van der Waals Attraction

The weakest type of interaction is electrostatic in nature and is known as van der Waals attraction. This type of attraction occurs between the nonpolar portion of two molecules and is brought about by a mutual distortion of electron clouds making up the covalent bonds (Fig. 1–1). This attraction is also called the induced dipole

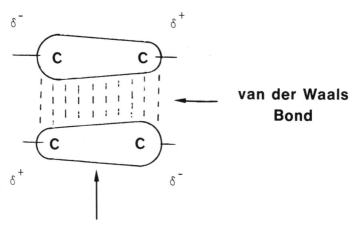

Fig. 1–1.   Van der Waal attraction resulting from distortion of covalent bonds

interaction. In addition to being weak, it is temperature dependent, being important at low temperatures and of little significance at high temperatures. The attraction occurs only over a short distance, thus requiring a tight packing of molecules. Steric factors therefore strongly influence van der Waals attraction. This type of chemical bonding is most prevalent in hydrocarbon and aromatic systems. Van der Waals forces are approximately 0.5 to 1.0 kilocalories per mole for each atom involved. Van der Waals bonds are found in lipophilic solvents but are of little importance in water.

### 2. Dipole-Dipole Bonding (Hydrogen Bond)

A stronger and important form of chemical bonding is the dipole-dipole bond, a specific example of which is the hydrogen bond (Fig. 1–2). A dipole results from the unequal sharing of a pair of electrons making up a covalent bond. This occurs when the two atoms making

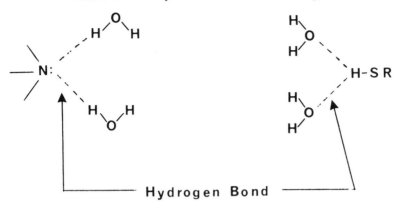

Fig. 1–2. Hydrogen bonding of an amine to water and a thiol to water

up the covalent bond differ significantly in electronegativity. A partial ionic character develops in this portion of the molecule, leading to a permanent dipole, with the compound being described as a polar compound. The dipole-dipole attraction between two polar molecules arises from the negative end of one dipole being electrostatically attracted to the positive end of the second dipole. The hydrogen bond can occur when at least one dipole contains an electropositive hydrogen (e.g., a hydrogen covalently bonded to an electronegative atom such as oxygen, sulfur, nitrogen, or selenium), which in turn is attracted to a region of high electron density. Atoms with high electron densities are those with unshared pairs of electrons such as amine nitrogens, ether or alcohol oxygens, and thioether or thiol sulfurs. While hydrogen bonding is an example of dipole-dipole bonding, not all dipole-dipole bonding is hydrogen bonding. Water, the important pharmaceutical solvent, is a good example of a hydrogen-bonding solvent. The ability of water to hydrogen bond accounts for the unexpectedly high boiling point of water as well as the characteristic dissolving properties of water. The hydrogen bond depends on temperature and distance. The energy of hydrogen bonding is 1.0 to 10.0 kcal/mole for each interaction.

### 3. Ionic Attraction

A third type of bonding is the ionic attraction found quite commonly in inorganic molecules and salts of organic molecules. This is formed by the attraction of a negative atom for a positive atom. The ionic bond involves a somewhat stronger attractive force of 5 kcal/mole or more and is *least* affected by temperature and distance.

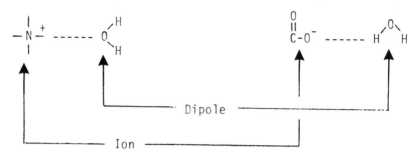

Fig. 1–3.   Ion-dipole bonding of a cationic amine to water and an anionic carboxylate to water

## 4. Ion-Dipole Bonding

Probably one of the most important chemical bonds involved in organic salts dissolving in water is the ion-dipole bond (Fig. 1–3). This bond occurs between an ion, either cation or anion, and a formal dipole, such as is found in water. The following two types of interactions may exist.

a. A cation will show bonding to a region of high electron density in a dipole (e.g., the oxygen atom in water).
b. An anion will bond to an electron-deficient region in a dipole (e.g., the hydrogen atom in water).

The ion-dipole is a strong attraction that is relatively insensitive to temperature or distance. When an organic compound with basic properties (e.g., an amine) is added to an aqueous acidic medium (pH below 7.0), the compound may form an ionic salt that, if dissociable, will have enhanced water solubility owing to ion-dipole bonding. Likewise, when an organic compound with acidic properties (e.g., carboxylic acids, phenols, unsubstituted or monosubstituted sulfonamides, and unsubstituted imides) is added to an aqueous basic medium (pH above 7.0), the compound may form an ionic salt, which, if dissociable, will have enhanced water solubility owing to ion-dipole bonding. Both of these examples are shown in Figure 1–3.

Water is an important solvent from both a pharmaceutical and a biologic standpoint. Therefore, when looking at any drug from a structural viewpoint, it is important to know whether the drug will dissolve in water. To predict water solubility, one must weigh the number and strength of hydrophilic groups in a molecule against the lipophilic groups present. If a molecule has a large amount of water-loving character, by interacting with water through hydrogen bonding or ion-dipole attraction, it would be expected to dissolve in water. If a molecule is deficient in hydrophilic groups but instead has a lipophilic portion capable of van der Waals attraction, then the molecule will most likely dissolve in nonaqueous or lipophilic media.

In reviewing the functional groups in organic chemistry, an attempt will be made to identify the lipophilic or hydrophilic character of each functional group. Knowing the character of each functional group in a drug will then allow an intelligent prediction of the overall solubility of the molecule by weighing the importance of each type of interaction. This book is organized in such a way that each functional group is discussed individually. Yet, when dealing with a drug molecule, the student will usually find a polyfunctional molecule. The ultimate goal is that the student should be able to predict the solubility of the actual drugs in water, aqueous acidic media, and aqueous basic media. Therefore, to use this book correctly and to prepare yourself for the drug molecules, it is recommended that you read through Chapter 16 after studying each functional group. This will help you put each functional group into perspective with respect to polyfunctional molecules.

# 2
# Alkanes $(C_nH_{2n+2})$

*A. Nomenclature.* The nomenclature of the alkanes may be either common or official nomenclature. The common nomenclature begins with the simplest system, methane, and proceeds to ethane, propane, butane, and so forth. The "ane" suffix indicates that the

| | |
|---|---|
| $CH_4$ | Methane |
| $CH_3CH_3$ | Ethane |
| $CH_3CH_2CH_3$ | Propane |
| $CH_3CH_2CH_2CH_3$ | n-Butane |
| $CH_3\underset{\underset{CH_3}{\mid}}{CH}CH_3$ | iso-Butane |

molecule is an alkane. This nomenclature works quite well until isomeric forms of the molecule appear (e.g., molecules with the same empirical formulas but different structural formulas). In butane, there are only two ways to put the molecule together, but as we consider larger molecules, many isomers are possible, and the nomenclature becomes unwieldy. Thus, a more systematic form of nomenclature is necessary. The IUPAC (International Union of Pure and Applied Chemistry) nomenclature is the official nomenclature.

IUPAC nomenclature requires that one find the longest continuous alkane chain. The name of this alkane chain becomes the base name. The chain is then numbered so as to provide the lowest possible numbers to the substituents. The number followed by the name of each substituent then precedes the base name of the straight-chain alkane. An example of naming an alkane according to IUPAC nomenclature is shown in Figure 2–1. The longest continuous chain is

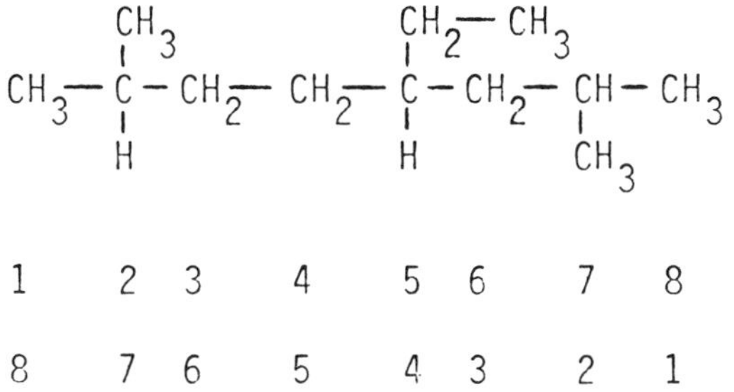

$$CH_3-\underset{\underset{H}{|}}{\overset{\overset{CH_3}{|}}{C}}-CH_2-CH_2-\underset{\underset{H}{|}}{\overset{\overset{CH_2-CH_3}{|}}{C}}-CH_2-\underset{\underset{CH_3}{|}}{CH}-CH_3$$

| 1 | 2 | 3 | 4 | 5 | 6 | 7 | 8 |

| 8 | 7 | 6 | 5 | 4 | 3 | 2 | 1 |

Fig. 2–1. 2,7-Dimethyl-4-ethyloctane

eight carbons. This chain can be numbered from either end. Numbering left to right results in substituents at positions 2 (methyl), 5 (ethyl), and 7 (methyl). The name of this compound would be 2,7-dimethyl-5-ethyloctane. Numbering from right to left gives alkane substituents at the 2, 4, and 7 positions. This compound would be 2,7-dimethyl-4-ethyloctane. To determine which way to number, just add the numbers that correspond to the substituent locations. From left to right, one has 2 + 5 + 7, which equals 14. When numbering from right to left, one has 2 + 4 + 7, which equals 13. Therefore, the correct numbering system is from right to left, giving 2,7-dimethyl-4-ethyloctane.

B. *Physical-Chemical Properties.* We wish to consider the following questions: Are alkanes going to be water soluble, and can water solubility or the lack of it be explained? The physical-chemical properties of alkanes are readily understandable from the previous discussion of chemical bonding. These compounds are unable to undergo hydrogen bonding, ionic bonding, or ion-dipole bonding. The only intermolecular bonding possible with these compounds is the weak van der Waal attraction. For the smaller molecules with one to four carbon atoms, this bonding is not strong enough to hold the molecules together at room temperature, with the result that the lower-member alkanes are gases. For the larger molecules with 5 to 20 carbon atoms, the induced dipole-induced dipole interactions can occur, and the energy required to break the increased amount of bonding is more than is available at room temperature. The result is that the 5- to 20-carbon atom alkanes are liquids. One can see from Table 2–1 that the boiling point increases consistently as more van der Waals bonding occurs.

The effects of adding an alkane to water are depicted in Figure 2–2. Water is an ordered medium with a considerable amount of intermolecular bonding, indicated by its high boiling point (i.e., high in respect to its molecular weight). In order to dissolve in or to mix with water, foreign atoms must break into this lattice. Sodium chloride (table salt), which is quite water soluble, is an example of a

**Table 2–1.**
Boiling Points of Common Alkanes

|  | Boiling Point($^{0}$C) |
|---|---|
| Propane ........................ | -42.0 |
| n-Butane ....................... | -0.5 |
| n-Pentane....................... | 36.1 |
| n-Hexane ....................... | 69.0 |
| n-Heptane....................... | 98.4 |
| n-Octane ....................... | 126.0 |

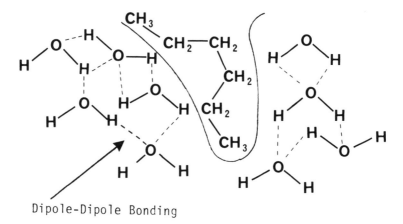

Fig. 2–2. Diagram of n-hexane's lack of solubility in water and the solubility of sodium chloride in water through ion-dipole bonding

molecule capable of this. An alkane cannot break into the water lattice since it cannot bond to water. Ion-dipole interaction, which is possible for sodium chloride, is not possible for the alkane. Ionic bonding and hydrogen bonding between water and the alkane also are not possible. Van der Waals bonding between alkane and alkane is relatively strong with little or no van der Waals attraction between the water and the alkane. The net result is that the alkane separates out and is immiscible in water. Alkanes will dissolve in a lipid solvent or oil layer. The term lipid, fat, or oil, defined from the standpoint of solubility, means a water-immiscible or water-insoluble material. Lipid solvents are rich in alkane groups; therefore, it is not surprising that alkanes are soluble in lipid layers, since induced dipole-induced dipole bonding will be abundant. If an alkane has a choice between remaining in an aqueous area or moving to a lipid area, it will move to the lipid area. In chemistry, this means that if n-heptane is placed in a separatory funnel containing water and decane, the n-heptane will partition into the decane. This movement of alkanes also occurs in biologic systems and is best represented by the general anesthetic alkanes and their rapid partitioning into the lipid portion of the brain while at the same time they have poor affinity for the aqueous blood. This concept will be discussed in detail in courses in medicinal chemistry.

Another property that should be mentioned is chemical stability. In the case of alkanes, one is dealing with a stable compound. For our purposes, these compounds are to be considered chemically inert to the conditions met "on the shelf," namely, air, light, aqueous acid or base, and heat.

A final physical-chemical property that may be encountered in branched chain alkanes is seen when a carbon atom is substituted with four different substituents (Fig. 2–3). Such a molecule is said to be asymmetric (that is, without a plane or point of symmetry) and is referred to as a chiral molecule. Chirality in a molecule means that the molecule exists as two stereoisomers which are nonsuperimposable mirror images of each other, as shown in Figure 2–3. These stereoisomers are referred to as enantiomeric forms of the molecule and possess slightly different physical properties. In addition, chirality in a molecule usually leads to significant biological differences in biologically active molecules. The topic of stereoisomerism is briefly reviewed in Appendix A.

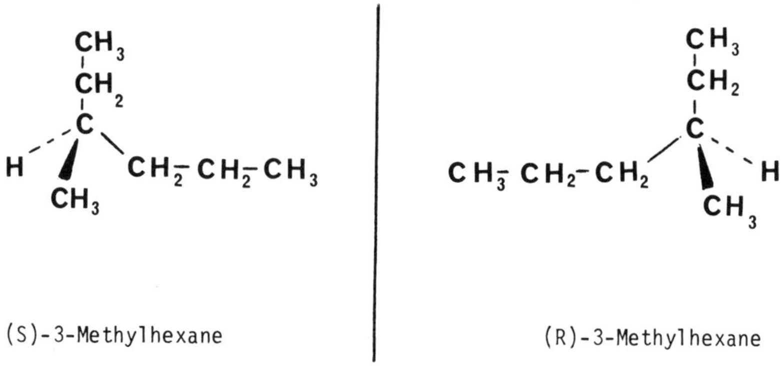

(S)-3-Methylhexane          (R)-3-Methylhexane

Fig. 2–3. Structures of (S)-3-methylhexane and its mirror image, (R)-3-methylhexane

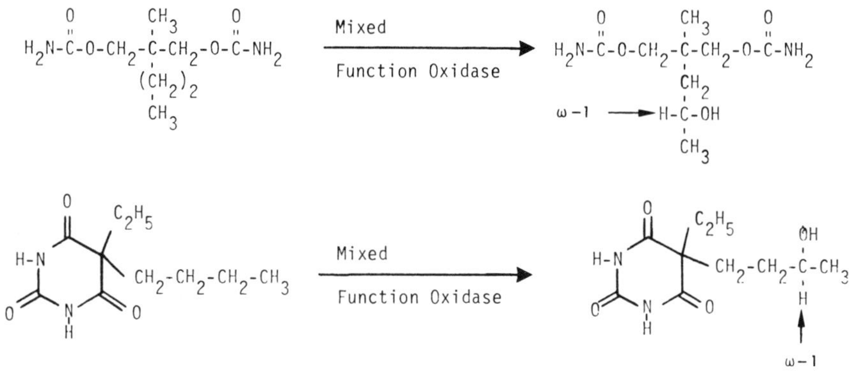

Fig. 2–4. Metabolism of meprobamate and butylbarbital by mixed function oxidase

C. *Metabolism.* The alkane functional group is relatively non-reactive in vivo and will be excreted from the body unchanged. Although the student should consider the alkanes themselves as nonreactive and the alkane portions of a drug as nonreactive, several notable exceptions will be emphasized in the medicinal chemistry courses, and they should be learned as exceptions. Two such exceptions are shown in Figure 2–4. When metabolism does occur, it is an oxidation reaction catalyzed by the mixed-function oxidase enzymes, and in most cases it occurs at the end of the hydrocarbon, the omega carbon, or adjacent to the final carbon at the omega-minus-one carbon, as shown.

# 3
# Alkenes ($C_nH_{2n}$)

| | |
|---|---|
| $CH_2 = CH_2$ | Ethyl<u>ene</u> |
| $CH_2 = CH\text{-}CH_3$ | Propyl<u>ene</u> |
| $CH_2 = CH\text{-}CH_2\text{-}CH_3$ | 1-Butyl<u>ene</u> |
| $CH_2 = \underset{\underset{CH_3}{\mid}}{C}\text{-}CH_3$ | iso-Butyl<u>ene</u> |

*A. Nomenclature.* The common nomenclature for the alkenes uses the radical name representing the total number of carbons present and the suffix, "ene," which indicates the presence of a double bond. This type of nomenclature becomes awkward for branched-chain alkenes, and the official IUPAC nomenclature becomes useful. In IUPAC nomenclature, the longest continuous chain containing the double bond is chosen and is given a base name that corresponds to the alkane of that length. In Figure 3–1, the longest chain has seven carbons and is therefore a heptane derivative. The chain is numbered so as to assign the lowest possible number to the double

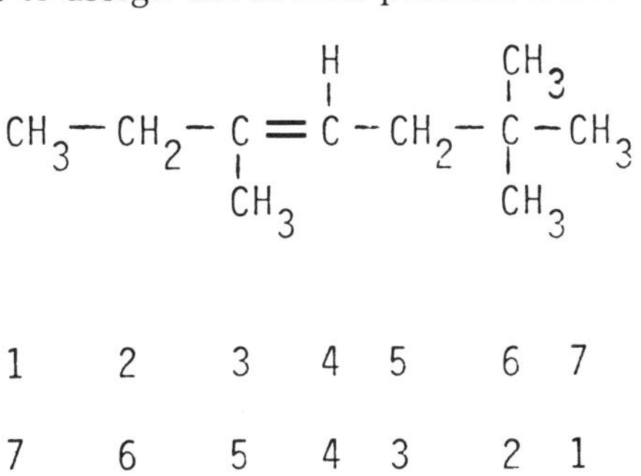

$$CH_3 - CH_2 - \underset{\underset{CH_3}{\mid}}{C} = \overset{\overset{H}{\mid}}{C} - CH_2 - \underset{\underset{CH_3}{\mid}}{\overset{\overset{CH_3}{\mid}}{C}} - CH_3$$

| 1 | 2 | 3 | 4 | 5 | 6 | 7 |
|---|---|---|---|---|---|---|
| 7 | 6 | 5 | 4 | 3 | 2 | 1 |

Fig. 3–1.  3,6,6-Trimethyl-3-heptene

11

bond. In numbering left to right, the double bond is at the 3 position, which is preferred, rather than numbering right to left, which would put the double bond at the 4 position. With the molecule correctly numbered, the final step in naming the compound consists of naming and numbering the alkyl radicals, followed by the location of the double bond and the alkane name, in which the "ane" is dropped and replaced with the "ene." In the example, the correct name would be 3,6,6,-trimethyl-3 (the location of the double bond) hept (meaning seven carbons) ene (meaning an alkene).

The introduction of a double bond into a molecule also raises the possibility of geometric isomers. Isomers are compounds with the same empirical formula but a different structural formula. If the difference in structural formulas comes from lack of free rotation around a bond, this is referred to as a geometric isomer. 2-Butene may exist as a *trans*-2-butene or *cis*-2-butene, which are examples of geometric isomers. Recently, the "E,Z" nomenclature has been insti-

$$CH_3 \diagdown \diagup H \qquad\qquad H \diagdown \diagup H$$
$$C = C \qquad\qquad C = C$$
$$H \diagup \diagdown CH_3 \qquad\qquad CH_3 \diagup \diagdown CH_3$$

trans 2-Butene                    cis 2-Butene

(E)-2-Butene                    (Z)-2-Butene

tuted to deal with tri- and tetrasubstituted alkenes, which cannot be readily named by *cis-trans* nomenclature. The "E" is taken from the German word *entgegen*, which means opposite, and the "Z" from *zusammen*, meaning together. Using a series of priority rules, if the two substituents of highest priority are on the same side of the $\pi$ bond, the configuration of Z is assigned, whereas if the two high-priority groups are on opposite sides, the E configuration is used. In the example in Figure 3–1, the correct nomenclature becomes (E)-3,6,6-trimethyl-3-heptene.

*B. Physical-Chemical Properties.*    The physical properties of the alkenes are similar to those of the alkanes. The lower members, having two through four carbon atoms, are gases at room temperature. Alkenes with five carbon atoms or more are liquids with increasing boiling points corresponding to increases in molecular weight. The weak intermolecular interaction that accounts for the low boiling point is again of the induced dipole-induced dipole type. Recognizing what type of intermolecular interaction is possible also allows a prediction of nonaqueous vs. aqueous solubility. Since alkenes cannot hydrogen bond and have a weak permanent

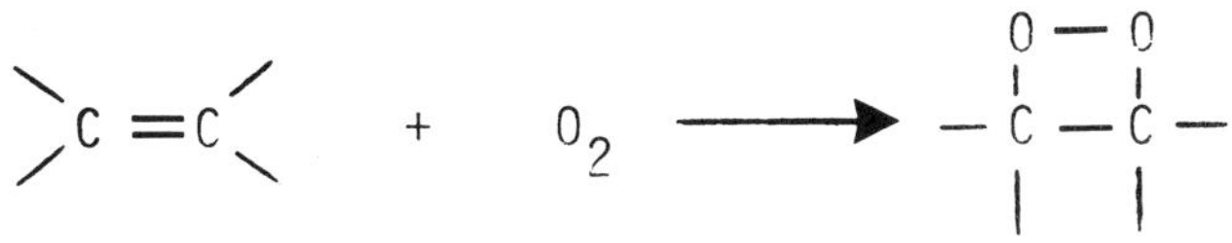

$$\text{C}=\text{C} \quad + \quad \text{O}_2 \quad \longrightarrow \quad -\overset{|}{\underset{|}{\text{C}}}-\overset{|}{\underset{|}{\text{C}}}-$$

Fig. 3–2. Oxidation of an alkene with molecular oxygen leading to a peroxide

dipole, they cannot dissolve in the aqueous layer. Alkenes will dissolve in nonpolar solvents such as lipids, fats, or oil layers. Therefore, physical properties of alkenes parallel those properties of the alkanes. When the chemical properties are considered, a departure from similarity to the alkane is found. The multiple bond gives the molecule a reactive site. From a pharmaceutical standpoint, alkenes are prone to oxidation, leading to peroxide formation (Fig. 3–2). Peroxides are quite unstable and may explode. In addition, alkenes, especially the volatile members, are quite flammable and, in the presence of oxygen and a spark, my explode.

C. *Metabolism.* Metabolism of the alkenes, as with the previously discussed alkanes, is not common. For our purposes, the al-

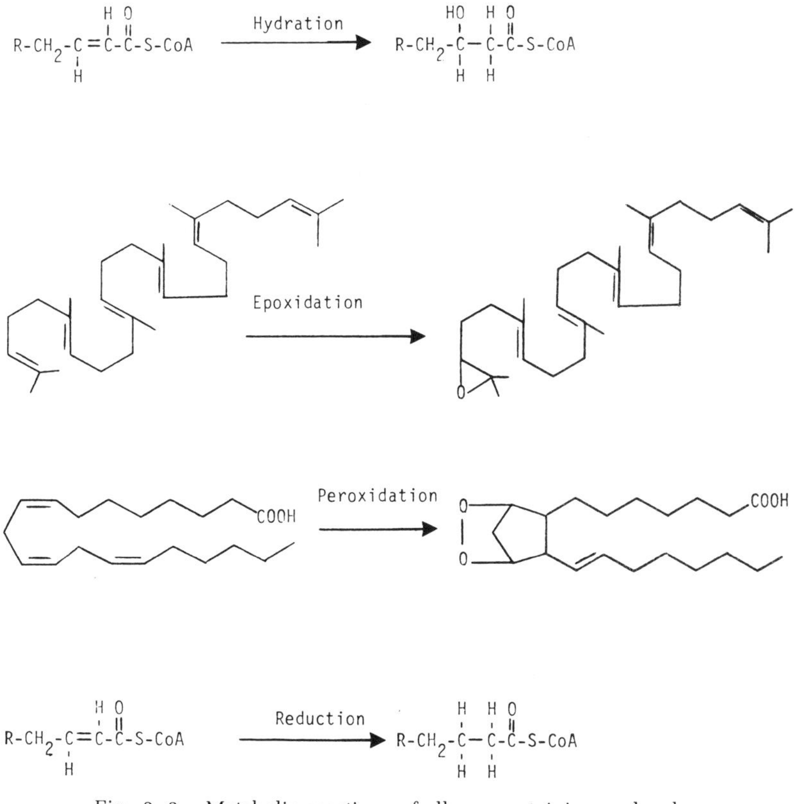

Fig. 3–3. Metabolic reactions of alkene-containing molecules

kene functional group should be considered metabolically stable. While alkene-containing drugs are usually stable in the body, the alkene functional groups of several body metabolites serve as centers of reaction. The unsaturated fatty acids add water to give alcohols. Mixed-function oxidase attacks the alkene functional group in squalene to give an epoxide during the biosynthesis of steroids. A peroxide intermediate is formed from eicosatrienoate, a triene, during prostaglandin biosynthesis, and during saturated fatty acid synthesis, alkenes are reduced in vivo (Fig. 3–3). You should be familiar therefore with these possible reactions of the alkene functional group and should not be surprised if an alkene-containing drug is metabolized.

Before leaving the topic of alkenes, a group of compounds that are isomeric to the alkenes should be mentioned. The cycloalkanes have the same empirical formula, $C_nH_{2n}$, as the alkenes but possess a different structural formula and are therefore isomeric. Three important members of this class are cyclopropane, cyclopentane, and

Cyclopropane          Cyclopentane          Cyclohexane

Reactive          Unreactive          Unreactive

Fig. 3–4. Common cyclic alkanes

cyclohexane (Fig. 3–4). Cyclopropane acts chemically like propene, while cyclopentane and cyclohexane are chemically inert, much like the alkanes. All three compounds are lipid soluble and quite flammable. The latter two ring systems are common to many drug molecules.

**QUESTIONS**

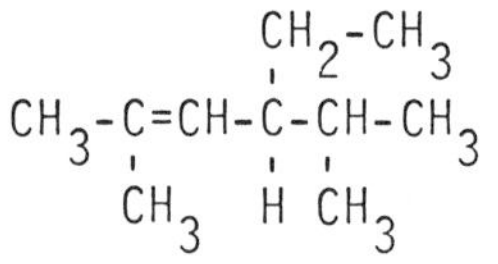

$$CH_3-C=CH-\overset{\overset{\displaystyle CH_2-CH_3}{|}}{\underset{\underset{\displaystyle CH_3}{|}}{C}}-\overset{\overset{}{}}{\underset{\underset{\displaystyle CH_3}{|}}{CH}}-CH_3$$

1. What is the IUPAC name of the compound shown?

   1. 2-Methyl-4-isopropyl-2-hexene
   2. 2,5-Dimethyl-3-ethyl-4-hexene
   3. 2,5-Dimethyl-4-ethyl-2-hexene
   4. 5-Methyl-3-(2-propyl) hex-4-ene
   5. 1,1,4-Trimethyl-3-ethyl-1-pentene

2. Which of the following systems would the compound dissolve in?

  1. Water
  2. Aqueous hydrochloric acid
  3. Aqueous sodium hydroxide
  4. Decane

3. What type of chemical instability (in vitro) might be predicted for this compound?

  1. Chemically nonreactive
  2. Peroxide formation
  3. Addition of water across the multiple bond
  4. Epoxidation
  5. $\omega$-Oxidation

# 4

# Aromatic Hydrocarbons

*A. Nomenclature.* Another class of hydrocarbons, shown in the following illustration, is the aromatic hydrocarbons. In aromatic nomenclature, a single name is used for the aromatic nucleus. Several of the most common nuclei have been listed along with their official name and numbering system.

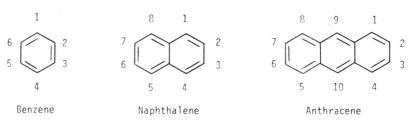

Benzene        Naphthalene        Anthracene

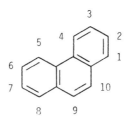

Phenanthrene

*B. Physical-Chemical Properties.* At first glance, it might be thought that the aromatic hydrocarbons are nothing more than cyclic alkenes, but this is not the case. Remember that the aromatic compounds do not have isolated single and double bonds, but instead they have a cloud of electrons above and below the ring. This is a cloud of delocalized electrons that are not as readily available as the

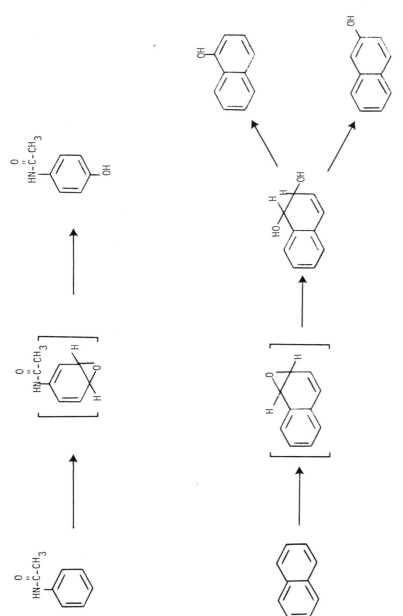

Fig. 4–1. Aromatic hydroxylation by liver microsomal enzymes

electrons in the alkenes. The aromatic systems are therefore not as prone to the chemical reactions that affect alkenes.

Formation of peroxides, a serious pharmaceutical problem with many alkenes, is not considered a problem with the aromatic hydrocarbons. The typical reaction of the aromatic systems is the electrophilic reaction. In the electrophilic reaction, the electrophile, the electron-loving, positively charged species, attacks the electron-dense cloud of the aromatic ring. There is one significant electrophilic reaction that occurs only in biologic systems, and this is known as hydroxylation. This reaction is quite important during drug metabolism but does not occur in vitro. Aromatic hydrocarbons are quite stable on the shelf. These hydrocarbons, like other hydrocarbons, are lipophilic and flammable. Because of their high electron density and flat nature, however, aromatic hydrocarbons show a somewhat stronger capacity to bond through van der Waals attraction. Aromatic rings appear to play a significant role in the binding of a drug to biologic proteins, as will be seen in courses on medicinal chemistry.

*C. Metabolism.*    As already mentioned, aromatic rings are quite prone to oxidation in vivo or, more specifically, to aromatic hydroxylation. This reaction commonly occurs in the liver microsomal enzymes and may involve an initial epoxidation. In a few cases, this highly reactive epoxide has been isolated, but in most cases the epoxide rearranges to give the hydroxylation product, the phenol or dialcohol, as shown in Figure 4–1. The importance of this reaction is considerable. The hydroxylation significantly increases the water solubility of the aromatic system. In many cases, this results in a rapid removal of the chemical from the body, while, in a few cases, hydroxylation may actually increase the activity of the drug. An area of considerable importance has been a study of the role of hydroxylation of aromatic hydrocarbons and its relationship to the carcinogenic properties of aromatic hydrocarbons. Evidence suggests that the intermediate epoxides are responsible for this carcinogenic effect.

# 5

# Halogenated Hydrocarbons

| | |
|---|---|
| $CH_3F$ | Methylfluoride |
| $CH_3CH_2Cl$ | Ethylchloride |
| $CH_3CH_2CH_2Br$ | Propylbromide |
| $CH_3CH_2CH_2CH_2I$ | n-Butyliodide |
| $CH_2Cl_2$ | Methylene chloride |
| $CHCl_3$ | Chloroform |
| $CCl_4$ | Carbon tetrachloride |
| $ClCH_2CH_2Cl$ | Ethylene chloride |

*A. Nomenclature.* The common nomenclature for mono-substituted halogenated hydrocarbons consists of the name of the alkyl radical followed by the name of the halogen atom. Examples of this nomenclature along with the structures and names of several common polyhalogenated hydrocarbons are shown.

This nomenclature again becomes complicated as the branching of the hydrocarbon chain increases, and one therefore uses IUPAC nomenclature. The IUPAC nomenclature requires choosing the longest continuous hydrocarbon chain, followed by numbering of the chain so as to assign the lowest number to the halide. The compound is then named as a haloalkane. This is illustrated in Figure 5–1 for the 4-methyl-2-bromopentane.

*B. Physical-Chemical Properties.* The properties of the halogenated hydrocarbons are different from those of the hydrocarbons previously discussed. The monohaloalkanes have a permanent dipole

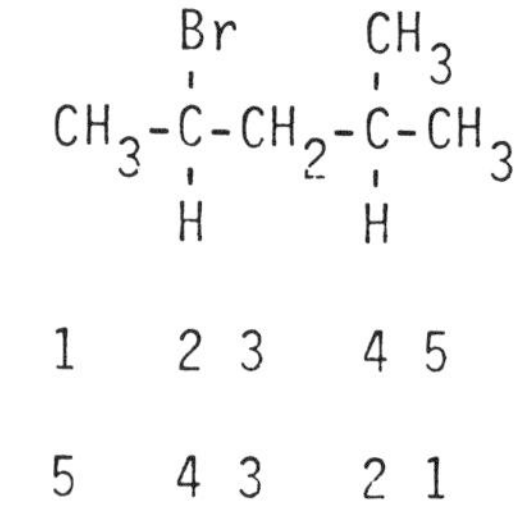

$$\begin{array}{ccc} & Br & CH_3 \\ & | & | \\ CH_3-C-CH_2-C-CH_3 \\ & | & | \\ & H & H \end{array}$$

1    2  3    4  5

5    4  3    2  1

Fig. 5–1.   4-Methyl-2-bromopentane

owing to the strongly electronegative halide attached to the carbon. The permanent dipole does not guarantee dipole-dipole bonding, however. Although the halogen is rich in electron density, there is no region highly deficient in electrons, and intermolecular bonding is therefore weak and again depends on the van der Waals attraction. Since only van der Waals bonding is possible, these compounds have low boiling points and poor water solubility. The halogens in general are thought to increase the lipophilic nature of the compounds. Another property of the halogenated hydrocarbons is a decrease in flammability with an increase in the number of halogens. In fact, carbon tetrachloride has been used in fire extinguishers. In general, these compounds are highly lipid soluble and chemically nonreactive.

One important chemical reaction that methylene chloride, chloroform, and several other polyhalogenated compounds undergo is shown in Figure 5–2. Chloroform, in the presence of oxygen and

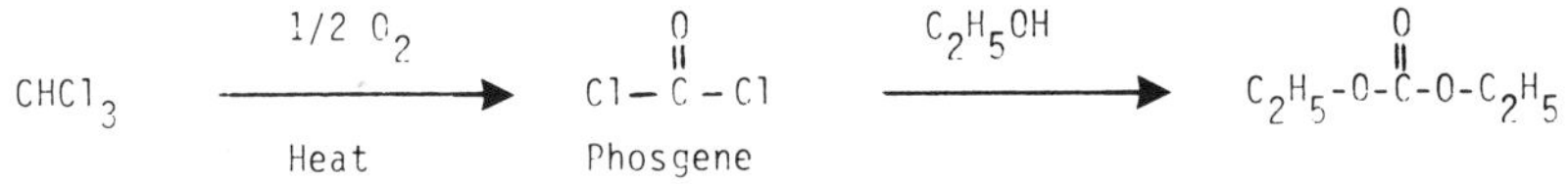

$$CHCl_3 \xrightarrow[\text{Heat}]{1/2\ O_2} \underset{\text{Phosgene}}{Cl-\overset{\overset{O}{\|}}{C}-Cl} \xrightarrow{C_2H_5OH} C_2H_5-O-\overset{\overset{O}{\|}}{C}-O-C_2H_5$$

Fig. 5–2.   Oxidation of chloroform to phosgene

heat, is converted to phosgene, a reactive and toxic chemical. To destroy any phosgene that may form in a bottle of chloroform, a small amount of alcohol is usually present. The alcohol reacts with the phosgene to give a nontoxic carbonate.

C. *Metabolism.*   The lack of chemical reactivity in vitro carries over to in vivo stability. In general, halogenated hydrocarbons are not readily metabolized. This stability significantly increases the potential for human toxicity. Since the compounds are quite lipid soluble, they are not readily excreted by the kidney. Since they are not rapidly metabolized to water-soluble agents, the halogenated hydrocarbons tend to have a prolonged biologic half-life, increasing the likelihood for systemic toxicity. This may also account for the potential carcinogenic properties of some halogenated hydrocarbons.

In summary, one significant property is common to all of the hydrocarbons, and that is the lack of ability to bond to water, and thus, the lipophilic or hydrophobic nature. SINCE ALL ORGANIC MOLECULES HAVE A HYDROCARBON PORTION, THIS PROPERTY WILL SHOW UP TO SOME EXTENT IN ALL MOLECULES. You will have to weigh the extent of influence of the lipophilic portion against the quantity of hydrophilic character in order to predict whether a molecule will dissolve in a nonaqueous medium or in water.

# 6
# Alcohols

-Common (Alkyl alcohol)

$CH_3OH$                Methyl alcohol (Wood alcohol)

$CH_3CH_2OH$            Ethyl alcohol (Alcohol U.S.P.)

$CH_3CHOH$             Isopropyl alcohol (Rubbing alcohol)
 |
 $CH_3$


      $CH_3$
       |
$CH_3C-OH$            tert-Butyl alcohol
       |
      $CH_3$


-IUPAC (Alkanol)

$CH_3OH$                Methanol

$CH_3CH_2OH$            Ethanol

$CH_3CHOH$             2-Propanol
 |
 $CH_3$


      $CH_3$
       |
$CH_3C-OH$            2-Methyl-2-propanol
       |
      $CH_3$

*A. Nomenclature.* The alcohols are named as "alcohol" preceded by the names of the hydrocarbon radical. Methyl and ethyl alcohol are examples of primary alcohols, isopropyl alcohol of a secondary alcohol, and tertiary butyl alcohol of a tertiary alcohol. The primary, secondary, and tertiary designation given to an alcohol depends upon the number of carbons that are attached to the carbon that contains the OH group. The primary designation indicates that one carbon is attached to the carbon bearing the OH group; the secondary designation that two carbons are attached, and the tertiary designation that three carbons are attached.

Once again, the nomenclature becomes clumsy as the hydrocarbon portion branches, and the official nomenclature must be used. The longest continuous chain that contains the hydroxyl group is chosen. The chain is then numbered to give the lowest number to the hydroxyl group. Other substituents preceded by their numbered location come first, followed by the location of the hydroxyl group, followed by the name of the alkane. To show that this is an alcohol, the "e" is dropped from the alkane name and replaced by "ol," the official sign of an alcohol.

B. *Physical-Chemical Properties.* The properties of the alcohol offer a departure from the compounds that have been discussed previously. The OH group can participate in intermolecular hydrogen bonding. Because of the electronegativity of the oxygen and the electropositive proton, a permanent dipole exists. The hydrogen attached to the oxygen is slightly positive in nature and the oxygen slightly negative. Remember, this is not a formal charge, but simply an unequal sharing of the pair of electrons that make up the covalent bond. The intermolecular hydrogen bonding that is now possible between the alcohol molecules results in relatively high boiling points as compared with their hydrocarbon counterparts (Table 6–1). Also important is the fact that the alcohol group can hydrogen bond to water. This means that it can break into the water lattice, with the result that the alcohol functional group promotes water solubility. The extent of water solubility for each alcohol will depend on the size of the hydrocarbon portion (Table 6–1). $C_1$ through $C_3$ alcohols are miscible with water in all proportions. As the length of the hydrocarbon chain increases, the hydrophilic nature of the molecule decreases. The location of the hydroxyl radical also influences water solubility, although not as dramatically as chain length. A hydroxyl group centered in the molecule will have a greater potential for producing water solubility than a hydroxyl at the end of the straight chain. If a second hydroxyl is added, solubility is increased. An example of this is 1,5-pentanediol. It can be thought of as ethanol and propanol put together. Since both alcohols are quite water soluble, it would be predicted that 1,5-pentanediol would also be quite water soluble, and it is. It also follows that, as the solubility of the alcohol in water decreases, the solubility of the alcohol in nonaqueous media increases. In summary, it can be said that an

**Table 6–1.**
Boiling Points and Water Solubility of Common Alcohols

|  | Boiling Points $^{0}C$ | Solubility (g /100g $H_2O$) |
|---|---|---|
| Methanol | 65.5 | ∞ |
| Ethanol | 78.3 | ∞ |
| 1-Propanol | 97.0 | ∞ |
| 2-Propanol | 82.4 | ∞ |
| 1-Butanol | 117.2 | 7.9 |
| 2-Butanol | 99.5 | 12.5 |
| 1-Pentanol | 137.3 | 2.3 |

alcohol functional group has the ability to solubilize to the extent of 1% or greater an alkane chain of five to six carbon atoms.

Looking at the chemical reactivity of the alcohol, it is found that, from a pharmaceutical standpoint, the alcohol functional group is a relatively stable unit. Remember, though, that in the presence of oxidizing agents, a primary alcohol will be oxidized to a carboxylic acid after passing through an intermediate aldehyde (Fig. 6–1).

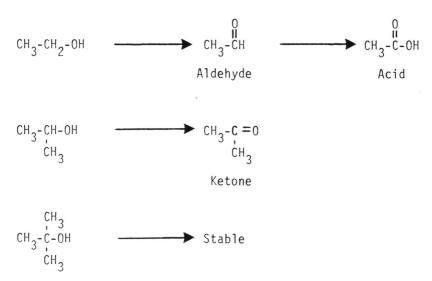

Fig. 6–1. Oxidation of a primary and secondary alcohol by oxidizing agents

The secondary alcohols can be oxidized to a ketone, and a tertiary alcohol is stable to mild oxidation. The oxidation of an alcohol in vitro is not commonly encountered because of the limited number of oxidating agents used pharmaceutically.

C. *Metabolism.* Although the alcohol functional group is relatively stable in vitro, it is readily metabolized in the body by a variety of enzymes. Both primary and secondary alcohols are prone to oxidation by oxidase enzymes resulting in formation of carboxylic acids or ketones, respectively (Fig. 6–2). The tertiary alcohols are

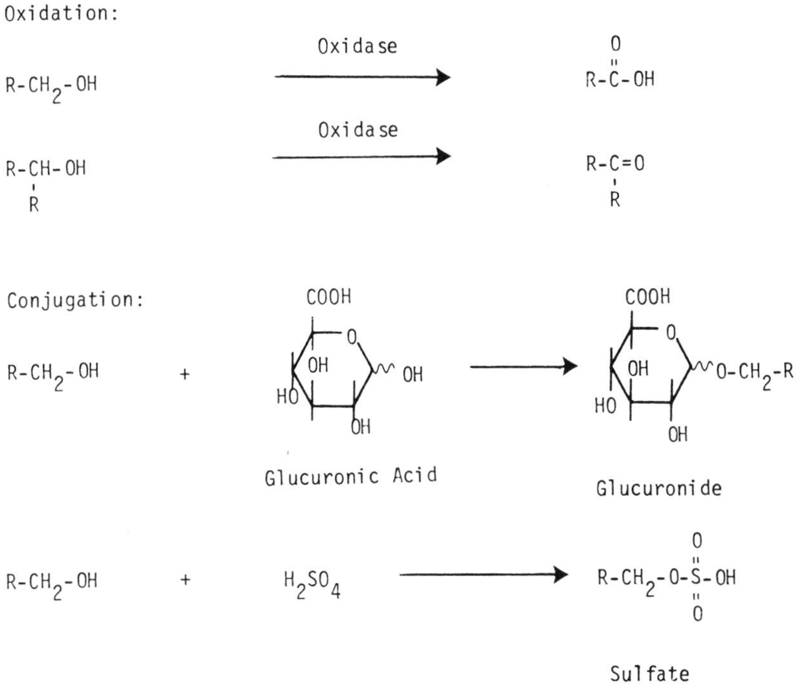

Oxidation:

$$R-CH_2-OH \xrightarrow{\text{Oxidase}} R-\overset{\overset{\displaystyle O}{\|}}{C}-OH$$

$$R-\underset{\underset{\displaystyle R}{|}}{CH}-OH \xrightarrow{\text{Oxidase}} R-\underset{\underset{\displaystyle R}{|}}{C}=O$$

Conjugation:

Glucuronic Acid → Glucuronide

$$R-CH_2-OH \;+\; H_2SO_4 \longrightarrow R-CH_2-O-\overset{\overset{\displaystyle O}{\|}}{\underset{\underset{\displaystyle O}{\|}}{S}}-OH$$

Sulfate

Fig. 6–2. Metabolic reactions of the alcohol functional group

stable to oxidase enzymes. Another common metabolic fate of the alcohol is conjugation with glucuronic acid to give a glucuronide or with sulfuric acid to give the sulfate conjugate. Both of these conjugates show a considerable increase in water solubility. The glucuronide has several additional alcohol functional groups that exhibit dipole-dipole bonding to water. The alcohol is conjugated to the glucuronic acid as an acetal through an ether-like linkage, while the conjugation to sulfuric acid is as an ester linkage.

When the alcohol combines with sulfuric acid, it is excreted as a sulfate conjugate, which would also be expected to show considerable water solubility because of the hydrogen bonding and ion-dipole bonding afforded by the sulfate portion of the molecule.

**QUESTIONS**

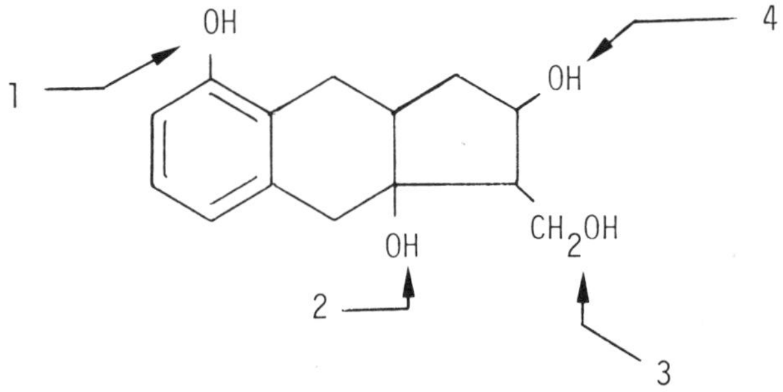

4. Which of the units shown in the compound is a tertiary alcohol?

   1. Alcohol 1
   2. Alcohol 2
   3. Alcohol 3
   4. Alcohol 4
   5. Alcohols 1 and 2

5. What property or properties would you predict for the compound?

   1. Inflammable
   2. Soluble in water
   3. Insoluble in water
   4. Soluble in aqueous acid

6. What type of instability would be expected for the alcohol portion of the molecule in the presence of an oxidizing agent?

   1. Alcohols 2, 3, and 4 would be oxidized to carboxylic acids.
   2. Alcohol 2 is stable, and alcohols 3 and 4 would be oxidized to ketones.
   3. Alcohol 2 is stable, while alcohol 3 would be oxidized to an acid and alcohol 4 to a ketone.
   4. All functional groups are stable to oxidizing agents.

# 7

# Phenols

A. *Nomenclature.* Phenols may appear to have some similarity to the alcohol functional group, but are considerably different in several aspects. Phenols differ from alcohols by having the OH group attached directly to an aromatic ring. The nomenclature of the phenols is not as systematic as has been the case with the previous functional groups. In many cases, phenols are named as substituted phenols using either the common ortho, meta, or para nomenclature for the location of the substituents, or the official nomenclature, in which the ring is numbered with the carbon that bears the OH having the 1 position. In phenol nomenclature, common names are often used, such as the cresol, catechol, and resorcinol. Therefore, one must be aware of these common names as well as the official nomenclature.

| Phenol | 4-Nitrophenol | o-Cresol | Catechol | Hydroquinone | Resorcinol |
| (Carbolic Acid) | (p-Nitrophenol) | | | | |

B. *Physical-Chemical Properties.* In considering the physical properties of the phenols, one is again aware of the OH group, in which a strong electronegative group, oxygen, is attached to the electropositive hydrogen. The permanent dipole is capable of intermolecular hydrogen bonding, which results in high boiling points and water solubility. Added to the list of compounds in Table 7–1 is cyclohexanol. Cyclohexanol differs from phenol only in the lack of the aromatic ring. The change in the boiling point and solubility in going from cyclohexanol to phenol may seem rather large, and in-

**Table 7-1.**

Boiling Points and Water Solubility of Common Phenols

|  | Boiling Point $^{O}C$ | Solubility (g/100g $H_2O$) |
|---|---|---|
| Cyclohexanol | 161 | 3.6 |
| Phenol | 182 | 9.3 |
| p-Cresol | 202 | 2.3 |
| m-Chlorophenol | 214 | 2.6 |
| Catechol | 246 | 45.0 |

deed should be, since a property becomes important with phenols that is absent in alcohols. That property is *acidity*.

Before discussing the acidity of the phenols, let us look at some additional factors that affect solubility. As the lipophilic nature of the phenol is increased, the water solubility is decreased. The addition of a methyl(cresol) or a halogen(chlorophenol) greatly reduces the water solubility of these compounds. The addition of a second hydroxyl, such as in catechol, increases water solubility, as was the case with the previous diols. The solubility of catechol will again greatly decrease as alkyls are added to this molecule.

The acidity of phenol and substituted phenols is considered in the following illustration. First, an acid must be defined. The classic definition states that an acid is a chemical that has the ability to give up a proton. Phenol has this ability and can therefore be considered an acid. The ease with which this proton is given up will influence the ratio of $K_1$ to $K_{-1}$. If $K_1$ is much greater than $K_{-1}$, a strong acid exists, while if $K_1$ is smaller than $K_{-1}$, a weak acid will result. The factor that influences the ratio of $K_1$ to $K_{-1}$ is the stability of the anion formed, in this case, the phenolate anion. It should be recalled that the phenolate anion can be stabilized by resonance, that is, the overlap of the pair of electrons on the oxygen with the delocalized cloud of electrons above and below the aromatic ring (Fig. 7-1). This is something an alcohol cannot do because an alcohol hydroxyl is not adjacent to an aromatic system, and resonance stabilization does not occur. Therefore, dissociation of the hydrogen from the oxygen is not possible in alcohols, and, by definition, the inability to give up a proton means that the alcohol is not acidic but neutral.

Acid (Definition)   HX  +  $H_2O$  $\underset{K_{-1}}{\overset{K_1}{\rightleftharpoons}}$  $H_3O^+$  +  $X^-$

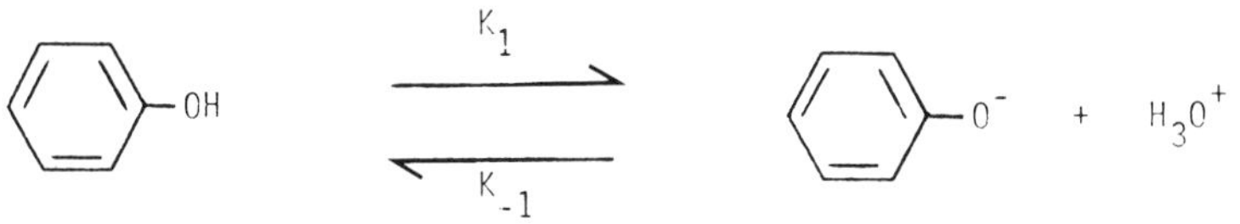

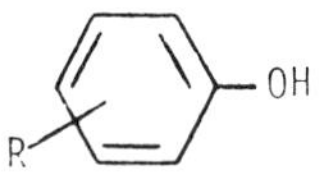

Dissociation constant ( Ka  and  pKa)

| | Ka (In water) | pKa |
|---|---|---|
| R = H | $1.1 \times 10^{-10}$ | 9.96 |
| R = m-CH$_3$ | $9.8 \times 10^{-11}$ | 10.01 |
| R = p-CH$_3$ | $6.7 \times 10^{-11}$ | 10.17 |
| R = m-NO$_2$ | $5.0 \times 10^{-9}$ | 8.3 |
| R = p-NO$_2$ | $6.9 \times 10^{-8}$ | 7.16 |
| Mineral Acids | $10^{-1}$ | 1.0 |
| Carboxylic Acids | $10^{-5}$ | 5.0 |
| Alcohols | $10^{-17}$ | 17.0 |

Fig. 7-1.   Dissociation constants in water for common phenols

Returning now to the question of boiling points and solubilities of alcohols vs. phenols, alcohols are neutral, and therefore the only type of intermolecular bonding is hydrogen bonding. This is not the only type of bonding between phenol and water. Phenols, which are acidic, exist both as neutral molecules and (to some extent) as ions; therefore, not only will hydrogen bonding occur but also the stronger ion-dipole bonding. The prediction of a higher boiling point and a greater water solubility relates to the presence of ion-dipole interaction as well as dipole-dipole bonding.

The acidity of phenols is influenced by the substitution on the aromatic ring. Substitution ortho to the phenol affects acidity in an unpredictable manner, while substitution meta or para to phenol results in acidities that are predictable. Substitution with a group capable of donating electrons onto the aromatic ring decreases acidity. The most pronounced effect occurs when the substitution is para or in direct conjugation. Addition of an electron-withdrawing group to the aromatic ring results in increased acidity. Again, the most pronounced effect occurs with para substitution. In both cases, the influence of substituents on acidity comes from the ability or inability of the substituent to stabilize the phenolate form. Comparison of the acidity of phenols to that of carboxylic acids and mineral acids shows that phenols are weak acids.

Another significant property of phenols is their chemical reactivity. An important reaction is shown in Figure 7–2. Because phenol is

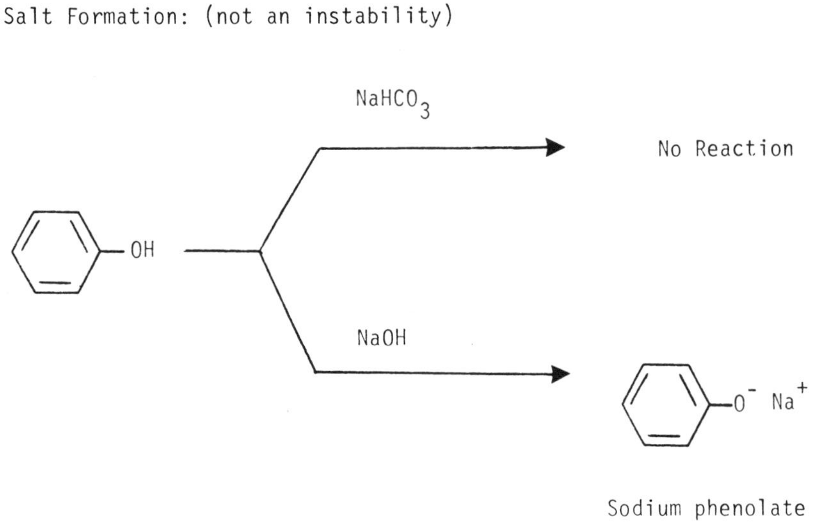

Salt Formation: (not an instability)

NaHCO$_3$ → No Reaction

NaOH → Sodium phenolate

Fig. 7–2. Acid-base reaction between phenol and a strong base

a weak acid, it will not react with sodium bicarbonate, a weak base, but will react with strong bases such as sodium hydroxide or potassium hydroxide to give the phenolate salts. Salt formation is an important reaction since the phenolates formed are ions and will dissolve in water through the much stronger ion-dipole bonding. As salts, the simple phenols (phenol, cresol, and chlorophenol) are extremely soluble in water. Several words of caution are necessary before leaving this topic. Sodium and potassium salts will greatly increase water solubility of the phenols. Heavy metal salts of the phenols will actually become less water soluble because of the inability of the salt to dissociate. Salts of phenols that are capable of dissociation will always increase water solubility, and for most of the phenols of medicinal value, the salts will give enough solubility

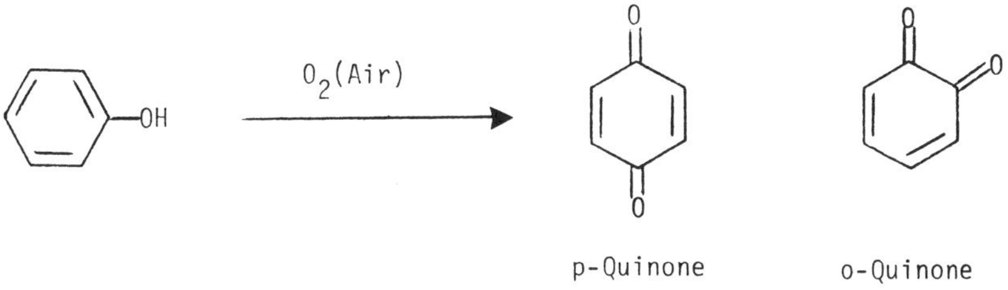

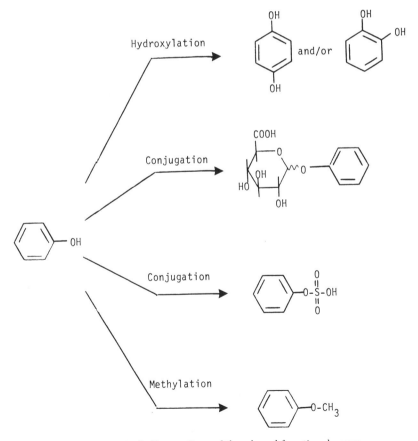

Fig. 7–3. Oxidation of phenol with molecular oxygen

so that the drug will dissolve in water at the concentration needed for biologic activity. As the lipophilic portions attached to the aromatic ring increase, however, the solubility of the phenolate salts will decrease. YOU SHOULD REALIZE THAT, WHILE SALT FORMATION (with a dissociating salt) IS AN EXAMPLE OF A CHEMICAL REACTION, IT IS NOT A CHEMICAL INSTABILITY. Treatment of the water soluble salt with acid will reverse this reaction, re-

Fig. 7–4. Metabolic reactions of the phenol functional group

generating the phenol. For our purposes, salt formation resulting in precipitation of the organic molecule is a pharmaceutical incompatibility that the student should watch for.

A second significant chemical reaction of phenols involves their facile air oxidation. Phenols are oxidized to quinones, which are highly colored. A clear solution of phenol allowed to stand in contact with air or light soon develops a yellow coloration owing to the formation of p-quinone or o-quinone (Fig. 7–3). This reaction occurs more readily with salts of phenols and with polyphenolic compounds. Phenols and their salts must be protected from oxygen and light by being stored in closed, amber containers or by the addition of antioxidants.

*C. Metabolism.* The metabolism of phenols is much like that of alcohols. The phenol may be oxidized, or, using the terminology previously used for aromatic oxidation, the phenol may be hydroxylated, to give a diphenolic compound (Fig. 7–4). In most cases, the new OH group will be either ortho or para to the original hydroxyl group. The most common form of metabolism of phenols is conjugation with glucuronic acid to form the glucuronide or sulfonation to give the sulfate conjugate. Both conjugation reactions give metabolites that have greater water solubility than the unmetabolized phenol. An additional type of metabolism seen to a minor extent is methylation of the phenol to give the methyl ether. This type of reaction will actually decrease water solubility.

**QUESTIONS**

7. What types of chemical bonds are possible between water and the compound shown?
   1. Van der Waals bonds
   2. Hydrogen bonds
   3. Ion-ion bonds
   4. Ion-dipole bonds
   5. Dipole-dipole bonds

   Predict the solubility of the compound in the media given: (Choose 1 for soluble or 5 for insoluble)

8. Water

9. Aqueous sulfuric acid

10. Aqueous potassium hydroxide

# 8

## Ethers

-Common (Alkylalkylether)

$CH_3-O-CH_2-CH_3$                    Ethylmethylether

$CH_3-CH_2-O-CH_2-CH_3$               Diethylether (Ether U.S.P.)

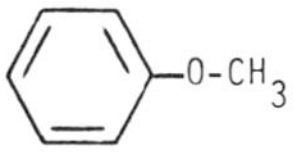

—$O-CH_3$        Methylphenylether (Anisole)

-IUPAC

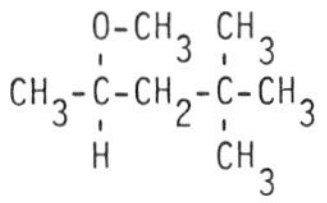

$$CH_3-\underset{\underset{H}{|}}{\overset{\overset{O-CH_3}{|}}{C}}-CH_2-\underset{\underset{CH_3}{|}}{\overset{\overset{CH_3}{|}}{C}}-CH_3$$

1   2 3   4 5              4,4-Dimethyl-2-methoxypentane (Correct)

5   4 3   2 1              2,2-Dimethyl-4-methoxypentane (Incorrect)

*A. Nomenclature.* Another important functional group found in many medicinal agents is the ether moiety shown in the above illustration. The ethers use a common nomenclature in which the compounds are called ethers, and both substituents are named by their radical names, such as methyl, ethyl, or phenyl. Thus, Ether USP, a common name, can also be referred to as diethylether. The inherent problem of naming the alkyl radical again arises as branching in the alkyl chain occurs. The official nomenclature names the compounds as alkoxy derivatives of alkanes. The longest continuous alkane chain containing the ether is chosen as the base name, and the alkane is numbered to give the ether the lowest number. The correct name for the ether shown is therefore 4,4-dimethyl-2-methoxy (numbered to give the alkoxy the lowest number) pentane (the longest alkane chain).

B. *Physical-Chemical Properties.*   What can one predict about the water solubility of the ether group? It is interesting that the synthesis of ethers is brought about by combining two alcohols or an alcohol and phenol to give the ether. The precursors have high boiling points, strongly bond to water to give solubility, and show chemical reactivity under certain conditions. Ethers, by contrast, are low-boiling liquids with poor water solubility (Fig. 8–1) and chemi-

Solubility

(g/100g $H_2O$)

Diethylether                                  8.4

Diisopropylether                           0.002

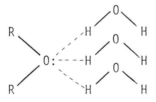

Fig. 8–1.   Diagrammatic representation of the solubility of an ether in water through dipole-dipole (hydrogen-bonding) bonding

cally are almost inert. This becomes understandable when one recalls that the properties of alcohols and phenols depend primarily upon the OH group. With diethyl ether, it can be seen that two molecules of ethanol are combined to give the diethyl ether, but during this process, the OH groups have been lost. Without the OH group, hydrogen bonds cannot exist, and the only intermolecular bonding is weak van der Waal attraction; thus, a low boiling point. Ether can hydrogen bond to water. The hydrogen of water will bond to the electron-rich oxygen (Fig. 8–1). The lower-membered ethers therefore show partial water solubility, but as the hydrocarbon portion increases, water solubility rapidly decreases. In the area of general anesthetics, this water solubility for ethers has a significant effect on the onset and duration of biologic activity. The figures given for water solubility for two ethers shown in Figure 8–1 demonstrate how rapidly water solubility decreases as the hydrocarbon portion increases.

Chemically the ethers are relatively nonreactive, stable entities with one important exception. Liquid ethers in contact with atmospheric oxygen form *peroxides* (Fig. 8–2). The peroxide formed,

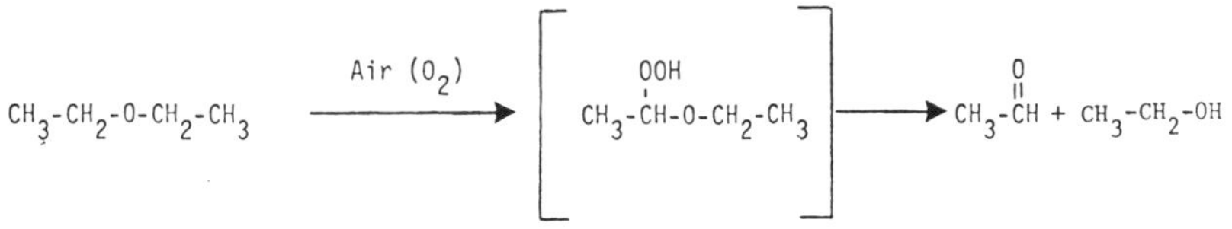

This reaction can be inhibited by copper metal

Fig. 8–2.  Oxidation of an ether with molecular oxygen to give a peroxide

although not present in great quantities, can be quite irritating to the mucous membranes and, if concentrated, may explode. Hence, care should be taken in handling ethers to minimize the contact with oxygen. Many times an antioxidant, such as copper metal, is added to take up any oxygen that may be present and thus prevent this instability.

C. *Metabolism.*    The metabolism of ethers in general is uneventful. With most ethers, one finds the ether excreted unchanged. There are exceptions to this rule, and the one exception that should be learned is the metabolic dealkylation reaction. When this does occur, the alkyl group that is lost is usually a small group such as a methyl or ethyl group. In the most common cases of dealkylation of an ether, a phenol forms, which is then metabolized by the routes of metabolism open to phenol, namely, the glucuronide or sulfate conjugation. The alkyl group is lost as an aldehyde: either formaldehyde, as shown in Figure 8–3, or acetaldehyde if the alkyl radical is ethyl.

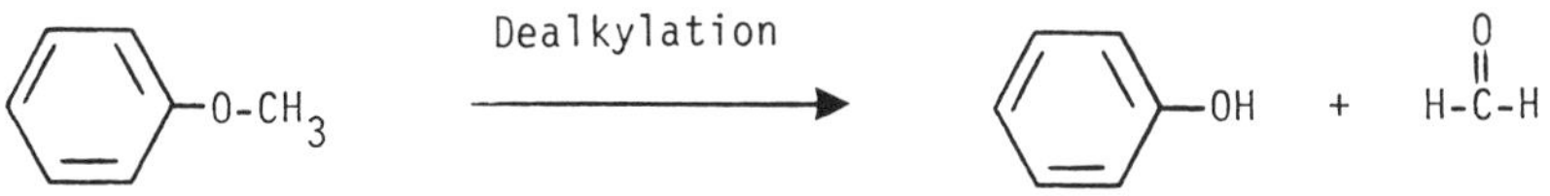

Fig. 8–3.  Metabolic dealkylation of anisole

**QUESTIONS**

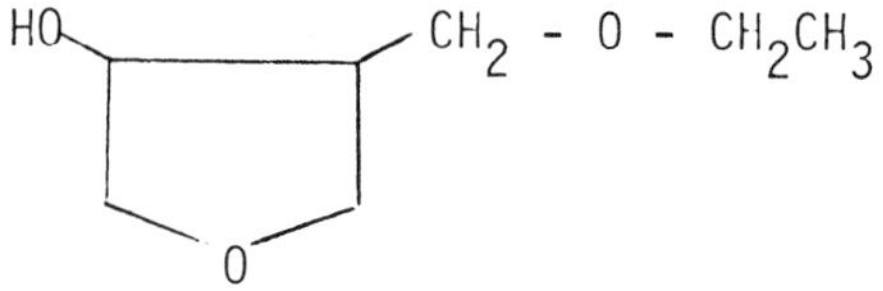

$$HO-\!\!\!-\!\!\!-\!\!\!-\!\!\!-\!\!\!-CH_2 - O - CH_2CH_3$$

(ring structure with O at the bottom)

Would you predict the compound shown to be soluble in:

11. Water?

   1. Yes
   2. No

12. Aqueous hydrochloric acid?

   1. Yes
   2. No

13. What in vitro (on-the-shelf) instability might be expected for this compound?

   1. Peroxide formation in the presence of air
   2. Aromatic oxidation
   3. Oxidation in the presence of air
   4. Hydrolysis
   5. Stable

# 9

# Aldehydes and Ketones

-Common

$$\underset{\text{Formaldehyde}}{\overset{\overset{\displaystyle O}{\|}}{H-C-H}}\qquad\qquad \underset{\text{Acetaldehyde}}{\overset{\overset{\displaystyle O}{\|}}{CH_3-C-H}}\qquad\qquad \underset{\text{Propionaldehyde}}{\overset{\overset{\displaystyle O}{\|}}{CH_3-CH_2-C-H}}$$

-IUPAC

$$\overset{\displaystyle CH_3}{\underset{\displaystyle CH_3}{CH_3-\overset{|}{\underset{|}{C}}-CH_2-\overset{\overset{\displaystyle O}{\|}}{C}-H}}\qquad\qquad \text{3,3-Dimethylbutanal}$$

$$\qquad\qquad\qquad\qquad\qquad\qquad -al = aldehyde$$

4   3 2   1

A. *Nomenclature.* Two functional groups that, owing to their chemical and physical similarities can be grouped together, are the aldehydes and ketones. From the examples of common names, it will be recognized that the word aldehyde appears as part of the name, formaldehyde for the one-carbon aldehyde, acetaldehyde for a two-carbon aldehyde, and propionaldehyde for the three-carbon aldehyde. The common nomenclature remains useful until we are unable to name the alkyl radical that contains the carbonyl, and then the formal IUPAC name is used. The longest continuous chain containing the aldehyde functional group is chosen as the base name and numbered such that the aldehyde constitutes the 1 position. To show the presence of an aldehyde in the molecule, the suffix "al" replaces the "e" in the alkane base name. Hence, the compound shown becomes 3,3-dimethylbutanal.

-Common

$$CH_3-\overset{\overset{\textstyle O}{\|}}{C}-CH_3 \qquad CH_3-\overset{\overset{\textstyle O}{\|}}{C}-CH_2-CH_3$$

Dimethyl<u>ketone</u>            Methylethyl<u>ketone</u>            Methylphenyl<u>ketone</u>

(Acetone)                                                    (Acetophenone)

-IUPAC

$$CH_3-\overset{\overset{\textstyle CH_3}{|}}{\underset{\underset{\textstyle H}{|}}{C}}-CH_2-\overset{\overset{\textstyle O}{\|}}{C}-\overset{\overset{\textstyle CH_3}{|}}{\underset{\underset{\textstyle H}{|}}{C}}-CH_2-\overset{\overset{\textstyle CH_3}{|}}{\underset{\underset{\textstyle CH_3}{|}}{C}}-CH_3$$

1    2 3    4 5 6    7 8

2,5,7,7-Tetramethyl-4-octan<u>one</u>

-<u>one</u> = ket<u>one</u>

With the common nomenclature of ketones, the word ketone is used as part of the nomenclature. With ketones, the two substituents are named using their radical name. Acetone, a common name, may also be called dimethylketone, while acetophenone may be referred to as methylphenylketone or phenylmethylketone. The IUPAC rules for ketones require that one find the longest continuous carbon chain that contains the ketone and number so as to give the lowest number to the carbonyl group. If the ketone is at the same location from either end of the molecule, then the correct direction of numbering is the one that gives the lowest number to any remaining substituents. The designation used to show the presence of a ketone carbonyl is the suffix "one" which, along with the ketone location, replaces the "e" in the alkane base name. The example given on this page becomes 2,5,7,7-tetramethyl-4-(the location of the carbonyl) octan(specifying an 8-carbon chain)-one(the abbreviation for a ketone).

*B. Physical-Chemical Properties.* In considering the properties of aldehydes and ketones, it must be noted that the carbonyl group present in both molecules is polar, and hence the compounds are polar. Oxygen is more electronegative than carbon, and the cloud of electrons that makes up the carbon-oxygen double bond is therefore distorted toward the oxygen. In addition, ketones and, to a lesser extent, aldehydes may exist in equilibrium with the "enol" form (Fig. 9–1). This property and the polar nature of the carbonyl lead to

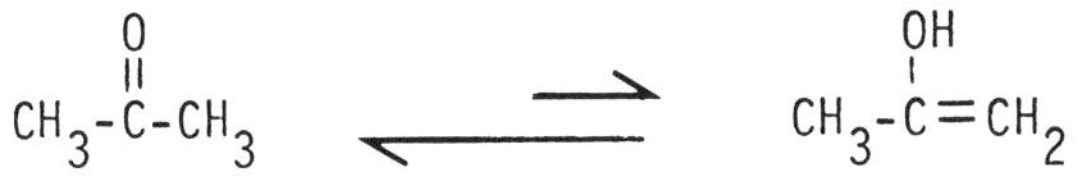

$$\underset{\text{CH}_3-\overset{\overset{\text{O}}{\|}}{\text{C}}-\text{CH}_3}{} \quad \rightleftharpoons \quad \underset{\text{CH}_3-\overset{\overset{\text{OH}}{|}}{\text{C}}=\text{CH}_2}{}$$

"Keto"                    "Enol"

Fig. 9–1. "Keto"-"enol" equilibrium of acetone

higher boiling points for aldehydes and ketones when compared with nonpolar compounds of comparable molecular weight. Because of the high electron density on the oxygen atom, aldehydes and ketones can hydrogen bond to water and will dissolve, to some extent, in water. The hydrogen bonding is similar to that suggested for ethers, but stonger. Keep in mind that as the nonpolar hydrocarbon portion increases, the effect of the polar carbonyl group on overall solubility will decrease. This is illustrated in Tables 9–1 and 9–2, where it is apparent that, as the hydrocarbon portion increases beyond two or three carbons, the water solubility decreases rapidly in both aldehydes and ketones. Some water solubility is still possible, however, with a total carbon content of five to six carbons.

In considering the chemical reactivity from a pharmaceutical standpoint, the ketone functional group is relatively nonreactive. This is not true of the aldehyde functional group. Aldehydes are one oxidation state from the stable carboxylic acid structure, and most are therefore rapidly oxidized. With many liquids, this means air oxidation, and compounds containing aldehydes therefore must be protected from atmospheric oxygen. The low-molecular-weight al-

**Table 9–1.**
Boiling Points and Water Solubility of Common Aldehydes

| $\text{R}-\overset{\overset{\text{O}}{\|}}{\text{CH}}$ | Boiling Point $^{\circ}$C | Solubility (g/100g $H_2O$) |
|---|---|---|
| R = H | −21 | ∞ |
| R = $CH_3$ | 20 | ∞ |
| R = $CH_3-CH_2$ | 49 | 16.0 |
| R = $CH_3-CH_2-CH_2$ | 76 | 7.0 |
| R = $C_6H_5$ | 178 | 0.3 |

**Table 9–2.**
Boiling Points and Water Solubility of Common Ketones

| $R-\overset{\overset{O}{\|}}{C}-R'$ | | Boiling Point $^{\circ}C$ | Solubility (g/100g $H_2O$) |
|---|---|---|---|
| R = $CH_3$ | R'= $CH_3$ | 56 | ∞ |
| R = $CH_3$ | R'= $CH_3-CH_2$ | 80 | 26.0 |
| R = $CH_3$ | R'= $CH_3-CH_2-CH_2$ | 102 | 6.3 |
| R = $CH_3-CH_2$ | R'= $CH_3-CH_2$ | 101 | 5.0 |
| R = $C_6H_5$ | R'= $CH_3$ | 202 | < 1.0 |

dehydes can also undergo polymerization to cyclic trimers, a compound containing three aldehyde units, or straight-chain polymers (Fig. 9–2). The trimers are stable to oxygen, but will allow regeneration of the aldehyde upon heating. In some cases, this reaction is used advantageously to protect the aldehyde.

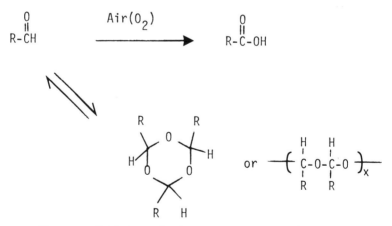

Fig. 9–2.   Oxidation and polymerization reactions of aldehydes

C. *Metabolism.*   Several possible metabolic routes are found in vivo for aldehydes and ketones. Aldehydes in general are readily oxidized by xanthine oxidase, aldehyde oxidase and by NAD-specific aldehyde dehydrogenase, the resulting product being a carboxylic acid (Fig. 9–3).

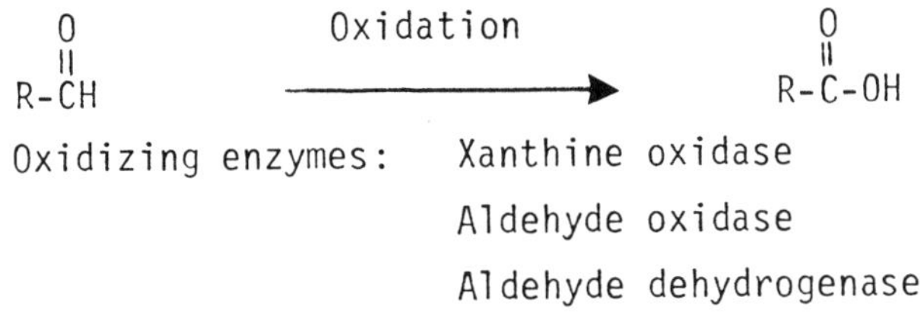

$$\underset{\text{R-CH}}{\overset{\overset{\text{O}}{\|}}{}} \xrightarrow{\text{Oxidation}} \underset{\text{R-C-OH}}{\overset{\overset{\text{O}}{\|}}{}}$$

Oxidizing enzymes:   Xanthine oxidase

Aldehyde oxidase

Aldehyde dehydrogenase

Fig. 9–3.   Metabolic oxidation of aldehydes

Ketones are fairly stable in vivo toward oxidation, although with an aromatic alkyl ketone, an oxidation reaction may occur to give an aromatic acid (Fig. 9–4).

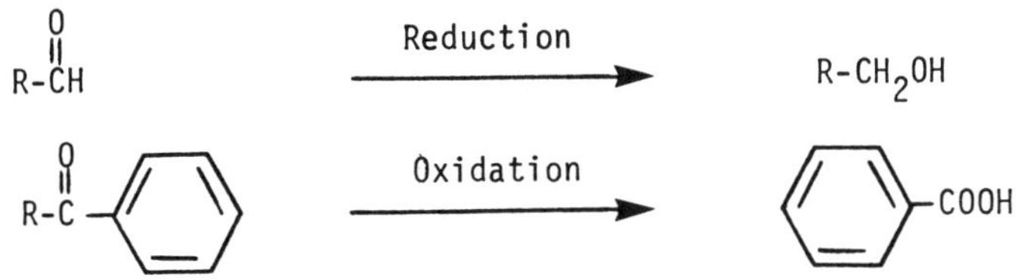

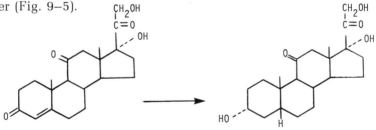

Fig. 9–4.   Minor metabolic reactions of aldehydes and ketones

A second metabolic reaction that may affect aldehydes and ketones is reduction. While reduction appears to be a minor metabolic reaction for aldehydes (Fig. 9–4), many ketones, especially $\alpha,\beta$-unsaturated ketones, undergo reduction to a secondary alcohol. This reaction is often stereoselective, giving rise primarily to one isomer (Fig. 9–5).

Fig. 9–5.   Metabolism of cortisone to tetrahydrocortisone

**QUESTIONS**

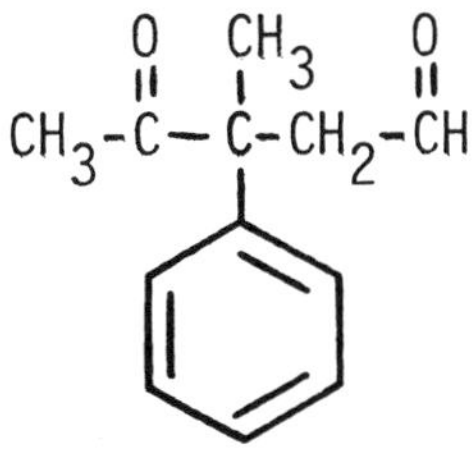

$$\underset{CH_3}{\overset{\overset{O}{\|}}{C}}-\underset{\underset{\text{(phenyl)}}{\overset{CH_3}{|}}}{C}-CH_2-\underset{}{\overset{\overset{O}{\|}}{CH}}$$

14.  What is the IUPAC name of the compound shown?

1.  $\beta$-Methyl-$\beta$-phenyl-$\gamma$-oxopentanal
2.  3-Methyl-4-oxo-3-phenylpentanal
3.  3-Methyl-3-phenyl-5-oxo-2-pentanone
4.  3-Methyl-3-acetylphenylacetaldehyde

15. Predict the in vitro instability of the compound.

    1. Oxidation of the ketone
    2. Oxidation of the aldehyde
    3. Alkyl oxidation
    4. Aromatic oxidation
    5. Peroxide formation

16. Predict the in vivo instability of the compound.

    1. Conjugation
    2. Aromatic hydroxylation
    3. Oxidation of the aldehyde
    4. Ketone oxidation
    5. Stable

# 10

# Amines

Two major functional groups still remain to be considered. These two groups, the carboxylic acids and the amines, are extremely important to medicinal chemistry and especially to the solubility nature of organic medicinals. In addition, the functional derivatives of these groups will be considered. In many instances the carboxylic acid or amine functional group is added to organic molecules with the specific purpose of promoting water solubility, since it is generally found that compounds showing little or no water solubility also are devoid of biologic activity.

-Common (Alkylamine)

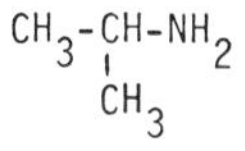

$CH_3-CH-NH_2$
  |
  $CH_3$                         Isopropyl<u>amine</u> (Primary amine)

$CH_3-CH_2-NH-CH_3$             Ethylmethyl<u>amine</u> (Secondary amine)

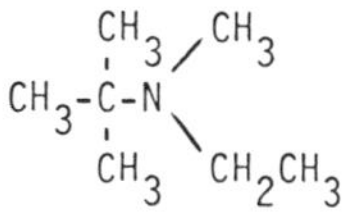

        $CH_3$  $CH_3$
          |   /
$CH_3-C-N$                      t-Butylethylmethyl<u>amine</u> (Tertiary amine)
          |   \
        $CH_3$  $CH_2CH_3$

-IUPAC

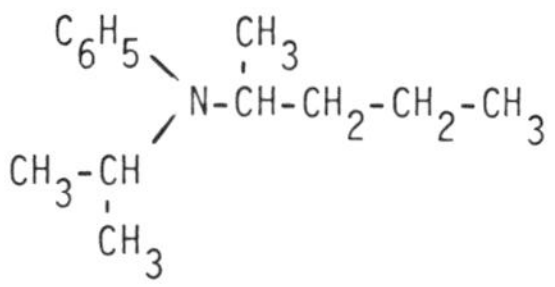

    $C_6H_5$    $CH_3$
        \        |
         / N-CH-CH_2-CH_2-CH_3   N-Phenyl-N-(2-propyl)-2-aminopentane
$CH_3-CH$
      |
      $CH_3$                     N= substituent on the <u>N</u>itrogen

43

A. *Nomenclature.* The common nomenclature for amines is illustrated on page 43. Inspection of this nomenclature reveals that the common names consist of the name of the alkyl or aryl radical, followed by the word amine. The examples given also show the different types of amines. The primary amine, isopropylamine, has a single substituent attached to the nitrogen; the secondary amine, methylethylamine, has two substituents attached to the nitrogen. The tertiary amine, t-butylmethylethylamine, has three groups attached directly to the nitrogen. As with all common nomenclatures, the system becomes nearly impossible to use as the branching of the alkyl groups increases, and the official nomenclature becomes necessary. In the IUPAC system, the amines are considered as substituted alkanes. The longest continuous alkyl chain containing the amine is identified and serves as the base name. The alkane chain is numbered in such a manner as to give the lowest possible number to the amine functional group, while the other substituents on the amine group are designated by use of a capital N before the name of the substituents. An example is given on page 43.

B. *Physical-Chemical Properties.* The amine functional group is probably one of the most common functional groups found in medicinal agents, and its value in the drug is twofold. One role is in solubilizing the drug either as the free base or as a water-soluble salt of the amine. The second role of the amine is to act as a binding site that holds the drug to a specific site in the body to produce the biologic activity. This latter role is beyond the scope of this book, but the former role contributes to an important physical property of the amine. First, let us again pose a question. What influence will the amine functional group have on solubility properties? While amines are polar compounds, they may not show high boiling points or good water solubility. One reason for this is that, in the tertiary amine, one does not find an electropositive group attached to the nitrogen. In the primary and secondary amines, one does have an electropositive hydrogen connected to the nitrogen, but the nitrogen is not as electronegative as oxygen, and the dipole is therefore weak. What all this means is that the amount of the intermolecular hydrogen bonding is minimal in primary and secondary amines and nonexistent in tertiary amines. This leads to relatively low-boiling liquids.

In considering water solubility, a different factor must be taken into account. The amine has an unshared pair of electrons, which leads to high electron density around the nitrogen. This high electron density promotes water solubility because hydrogen bonding between the hydrogen of water and the electron-dense nitrogen occurs. This is similar to the situation with low-molecular-weight ethers but occurs to a greater extent with basic amines. Both boiling

**Table 10–1.**
Boiling Points and Water Solubility of Common Amines

| $R_1$ | $R_2$ | $R_3$ | Boiling Point $^\circ$C | Solubility (g/100g $H_2O$) |
|-------|-------|-------|-------------------------|----------------------------|
| $CH_3$ | H | H | -7.5 | very soluble |
| $CH_3$ | $CH_3$ | H | 7.5 | very soluble |
| $CH_3$ | $CH_3$ | $CH_3$ | 3.0 | 91 |
| $C_2H_5$ | H | H | 17.0 | |
| $C_2H_5$ | $C_2H_5$ | H | 55.0 | very soluble |
| $C_2H_5$ | $C_2H_5$ | $C_2H_5$ | 89.0 | 14 |
| $C_6H_5$ | H | H | 184.0 | 3.7 |
| $C_6H_5$ | $CH_3$ | H | 196.0 | slightly soluble |
| $C_6H_5$ | $CH_3$ | $CH_3$ | 194 | 1.4 |

(Structure shown in header: $R_1$–N($R_3$)–$R_2$)

points and the solubility effects are shown in Table 10–1. Also illustrated in Table 10–1 is the effect on solubility of increasing the hydrocarbon portion. Primary amines tend to be more soluble than secondary amines, which are more soluble than tertiary amines. The amine can solubilize up to five or six methylenes, which, from a solubility standpoint, makes the amines equivalent to an alcohol.

An extremely important property of the amines is their basicity and ability to form salts. The Brønsted definition of a base is the ability of a compound to donate or share a pair of electrons. Amines have an unshared pair of electrons, which is more or less available for sharing. The statement "more or less" has to do with the strength of a base, and this is considered in Figure 10–1. The strength of a base is defined by its relative ability to donate its unshared pair of electrons. The more readily the electrons are donated, the stronger the base. Two factors influence the availability of the electrons. One of the factors is electronic, while the other is steric. To consider the former, if electron-donating groups are attached to the basic nitrogen, electrons are pushed into the nitrogen. Since a negative repels a negative, the electron pair on the nitrogen will be pushed out from the nitrogen, thus making them more readily available for donating.

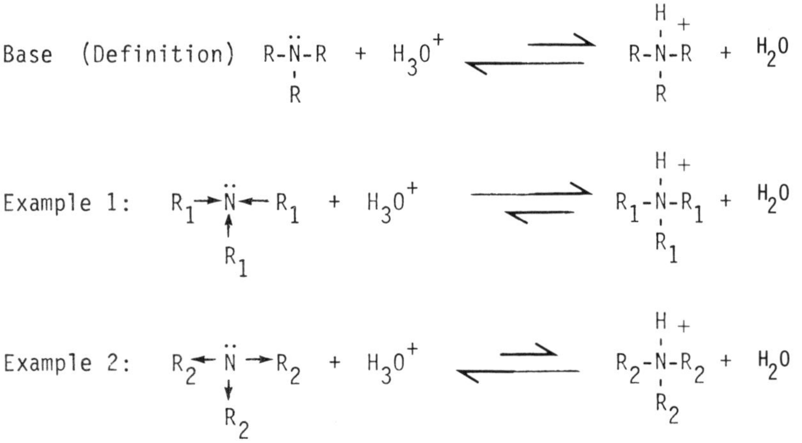

Base (Definition)   $R-\overset{..}{\underset{R}{N}}-R$ + $H_3O^+$ $\rightleftharpoons$ $R-\overset{H}{\underset{R}{N}}-R^+$ + $H_2O$

Example 1:   $R_1 \rightarrow \overset{..}{\underset{R_1}{N}} \leftarrow R_1$ + $H_3O^+$ $\rightleftharpoons$ $R_1-\overset{H}{\underset{R_1}{N}}-R_1^+$ + $H_2O$

Example 2:   $R_2 \rightarrow \overset{..}{\underset{R_2}{N}} \rightarrow R_2$ + $H_3O^+$ $\rightleftharpoons$ $R_2-\overset{H}{\underset{R_2}{N}}-R_2^+$ + $H_2O$

Fig. 10–1.   The influence of electron-releasing and electron-withdrawing groups on the basicity of amines

If, on the other hand, electron-withdrawing or electron-attracting groups are attached to the nitrogen, the unshared pair of electrons will be pulled to the nitrogen atom and will be less readily available for donating, and therefore a weaker base results. An example of the electron donor is the alkyl, and an example of an electron-withdrawing group is the aryl or phenyl group. Based on this, one would predict that secondary alkyl amines with two electron-releasing groups attached to the nitrogen should be more basic than primary alkyl amines with a single alkyl group attached to the nitrogen. This is normally true. One would also predict that tertiary alkyl amines with three electron-releasing groups attached to the nitrogen should be more basic than secondary amines. This would be true if it were not for steric hindrance, the second factor that affects basicity. If large alkyl groups surround the unshared pair of electrons, then the approach of hydronium ions, a source of a proton, is hindered.

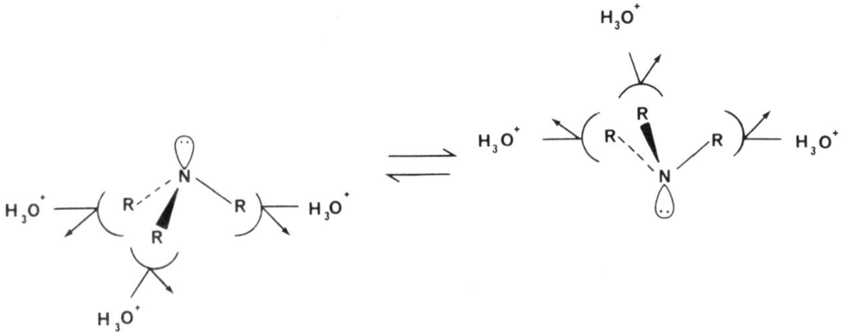

Fig. 10–2.   Diagrammatic representation of the influence of steric factors on the basicity of tertiary alkyl amines

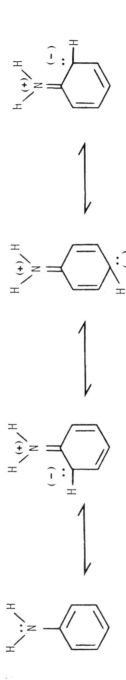

Fig. 10–3. Resonance stabilization of aniline's unshared electron pair

The degree of this hindrance will affect the strength of basicity. The steric effect becomes important for tertiary amines but has little, if any, effect on primary and secondary amines. As shown in Figure 10–2, with amines, the large alkyl groups move back and forth, blocking the approach of water. Salt formation therefore does not occur as readily as it would in the absence of such hindrance. We commonly find that with alkyl amines, secondary amines are more basic than tertiary amines, and tertiary amines are more basic than primary amines.

Aromatic amines differ significantly from alkyl amines in basicity. The aromatic ring, with its delocalized cloud of electrons, serves as an electron sink. The aromatic ring thus acts as an electron-withdrawing group, leading to a drop in basicity by six powers of ten. The unshared pair of electrons are said to be resonance stabilized, as shown in Figure 10–3. The spreading of the electron density over a greater area decreases the ability of the molecule to

**Table 10–2.**
Dissociation Constants and $pK_b$ Values in Water of
Common Amines

$$R_1-\underset{\overset{|}{R_2}}{N}-R_3$$

| $R_1$ | $R_2$ | $R_3$ | Dissociation constant ($K_b$ and $pK_b$) $K_b$ (In water) | $pK_b$ |
|---|---|---|---|---|
| $CH_3$ | H | H | $4.4 \times 10^{-4}$ | 3.36 |
| $CH_3$ | $CH_3$ | H | $5.1 \times 10^{-4}$ | 3.29 |
| $CH_3$ | $CH_3$ | $CH_3$ | $0.6 \times 10^{-4}$ | 4.22 |
| $CH_3(CH_2)_2$ | H | H | $3.8 \times 10^{-4}$ | 3.42 |
| $CH_3(CH_2)_2$ | $CH_3(CH_2)_2$ | H | $8.1 \times 10^{-4}$ | 3.09 |
| $CH_3(CH_2)_2$ | $CH_3(CH_2)_2$ | $CH_3(CH_2)_2$ | $4.5 \times 10^{-4}$ | 3.35 |
| $C_6H_5$ | H | H | $4.2 \times 10^{-10}$ | 9.38 |
| $C_6H_5$ | $CH_3$ | H | $7.1 \times 10^{-10}$ | 9.15 |
| $p-O_2N-C_6H_4$ | H | H | $1.0 \times 10^{-13}$ | 13.0 |
| $m-O_2N-C_6H_4$ | H | H | $3.2 \times 10^{-12}$ | 11.49 |
| $p-CH_3-C_6H_4$ | H | H | $1.2 \times 10^{-9}$ | 8.92 |
| $m-CH_3-C_6H_4$ | H | H | $4.9 \times 10^{-10}$ | 9.31 |
| $C_6H_5$ | $C_6H_5$ | H | $7.0 \times 10^{-14}$ | 13.15 |

donate the electrons, and basicity is therefore reduced. Additional substitution on the nitrogen of aniline with an alkyl or second aryl group changes the basicity in a predictable manner, with the alkyl group increasing basicity and an aryl reducing basicity to a nearly neutral compound (Table 10–2). Finally, substitution on the aromatic ring also affects basicity. Substitution meta or para to the amine has a predictable effect on basicity while ortho substitution affects basicity in an unpredictable manner (Table 10–2). An electron-withdrawing group attached to the aromatic ring in the meta or para position decreases basicity. The decrease is significant if this

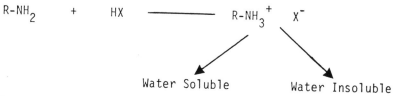

Fig. 10–4. The salt formed from an amine and an acid is water soluble if the salt is able to dissociate and is water insoluble if the salt is unable to dissociate

group is para rather than meta. Electron-donating groups in the meta or para position usually increase basicity above that of aniline. The increase in basicity is most pronounced if the group is in the para position and not as pronounced if it is in the meta position. It will be noted that this is just the opposite of phenols. With ortho-substituted anilines, predictability fails because of intramolecular interactions.

Since amines are basic, one would expect that they react with acids to form salts. This is an important reaction, for if the salts that are formed dissociate in water, there is a strong likelihood that these salts will be water soluble (Fig. 10–4). Such is the case with many organic drugs. If a basic amine is present in the drug, it can be converted into a salt, which in turn is used to prepare aqueous solutions of the drug. The most frequently used acids for preparing salts are hydrochloric, sulfuric, tartaric, succinic, citric, and maleic acids. Hydrochloric acid is a monobasic acid; it has one proton and therefore reacts with one molecule of base. The others are dibasic acids (sulfuric, tartaric, succinic, and maleic) and tribasic acids (citric and phosphoric). The aqueous solution of the amine salt will have a characteristic pH that will vary depending on the acid used. The pH will be acidic when a strong mineral acid is used to prepare the salt or weakly acidic or neutral if a weak organic acid is used. Since the amine is converted to a water-soluble salt by the action of the acid, it is reasonable to assume that the addition of a base to the salt would result in liberation of the free amine, which in

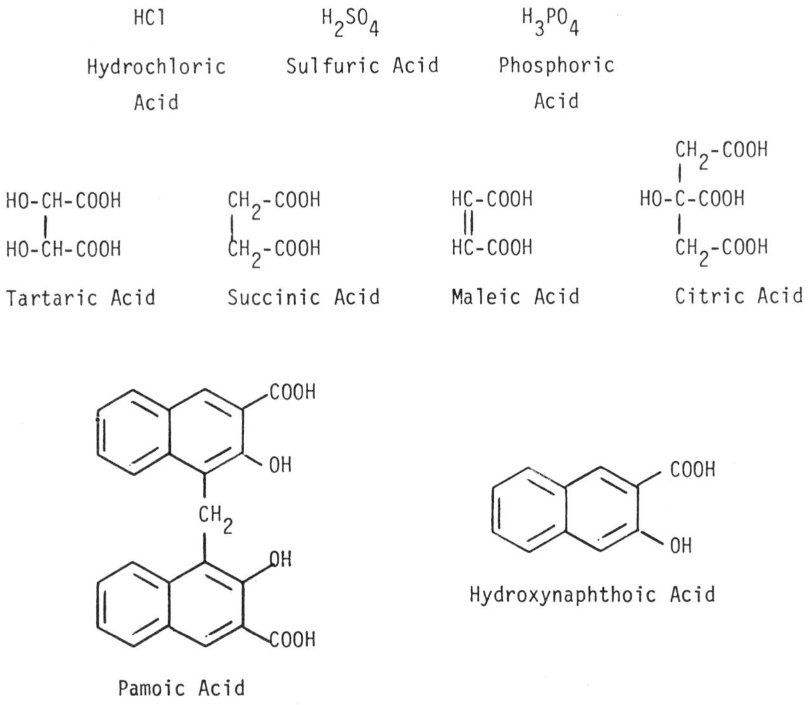

HCl                    $H_2SO_4$               $H_3PO_4$

Hydrochloric      Sulfuric Acid        Phosphoric
Acid                                              Acid

$$HO-CH-COOH$$
$$|$$
$$HO-CH-COOH$$

Tartaric Acid

$$CH_2-COOH$$
$$|$$
$$CH_2-COOH$$

Succinic Acid

$$HC-COOH$$
$$||$$
$$HC-COOH$$

Maleic Acid

$$CH_2-COOH$$
$$|$$
$$HO-C-COOH$$
$$|$$
$$CH_2-COOH$$

Citric Acid

Pamoic Acid

Hydroxynaphthoic Acid

Fig. 10–5.   Structures of common acids used to prepare salts of basic amines

turn may precipitate. This is a chemical incompatibility that could be quite important when drugs are mixed. Included in Figure 10–5 are two additional commonly used acids, pamoic and hydroxynaphthoic acid. These acids are commonly used in medicinal chemistry to form amine salts that are water insoluble, in other words, salts that will not dissociate. This property is used to good advantage in that it prevents a drug from being absorbed and thus keeps the drug in the intestinal tract.

   *C. Metabolism.*   Many metabolic routes are available for handling amines in the body, some of which are illustrated in Figure 10–6. A common reaction that secondary and tertiary amines undergo is dealkylation. In the dealkylation reaction, the alkyl group is lost as an aldehyde or ketone and the amine is converted from a tertiary amine to a secondary amine and finally to a primary amine. This reaction usually occurs when the amine is substituted with small alkyl groups such as a methyl, ethyl, or propyl group. An example of a drug metabolized by a dealkylation reaction is imipramine, which is metabolized to desimipramine. Primary alkyl amines can also undergo a dealkylation reaction of sorts, known as deamination. Here again, an aldehyde or ketone is formed along with an amine.

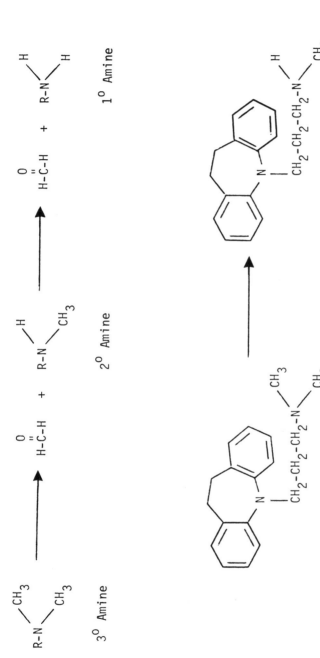

Fig. 10–6. Metabolic demethylation of tertiary and secondary amines

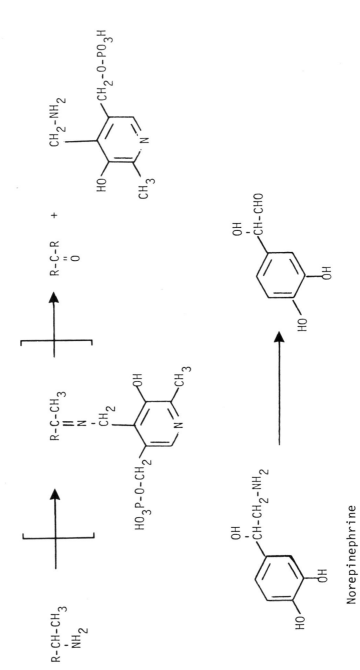

Fig. 10–7. Metabolic deamination of a primary amine catalyzed by pyridoxal phosphate

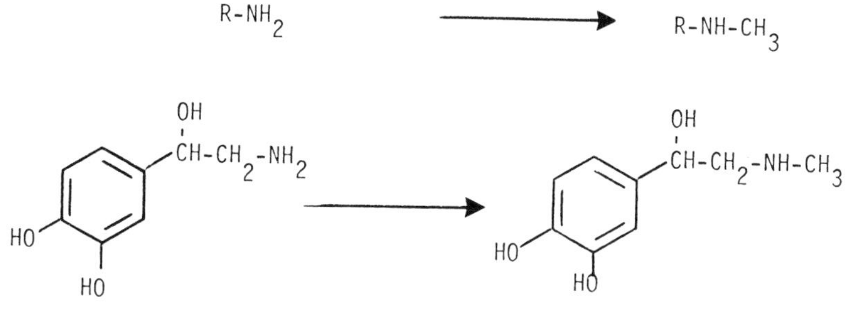

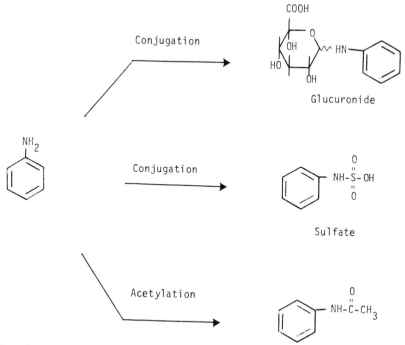

R-NH$_2$ $\longrightarrow$ R-NH-CH$_3$

Norepinephrine

Epinephrine

Fig. 10-8. Metabolic methylation of an amine

Pyridoxal 5-phosphate may catalyze this reaction, resulting in the formation of pyridoxamine. In order for this reaction to occur, a carbon bonded to the nitrogen must have at least one hydrogen. The enzymes most commonly found that catalyze deamination reactions are monoamine oxidase (MAO) and diamine oxidase (DAO). An example of a MAO-catalyzed reaction is the deamination of norepinephrine, as shown in Figure 10-7.

Conjugation

Glucuronide

Conjugation

Sulfate

Acetylation

Fig. 10-9. Metabolic conjugation of primary amines with glucuronic acid, sulfuric acid, or acetyl coenzyme A

A minor metabolic route open to amines is the methylation reaction. An important example of the methylation reaction is the biosynthesis of epinephrine from norepinephrine by the enzyme phenylethanolamine-N-methyltransferase (Fig. 10–8).

Far more important to the metabolism of primary and secondary, but not tertiary, amines are the conjugation reactions. Amines can be conjugated with glucuronic acid and sulfuric acid to give the glucuronides and sulfates, both of which exhibit a significant increase in water solubility. Amines, both primary and secondary, may also be acetylated by acetyl CoA to give a compound that usually shows a decrease in water solubility (Fig. 10–9).

## QUATERNARY AMMONIUM SALTS

Special amine derivatives with unique properties are the quaternary ammonium salts.

A. *Nomenclature.* While the reaction of primary, secondary, or tertiary amines with acid leads to the formation of the respective ammonium salts, these reactions can be reversed by treatment with base, regenerating the initial amines. The quaternary ammonium salts we wish to consider here are those compounds in which the nitrogen is bound to four carbon atoms through covalent bonds:

$1^0$ Amine     $R-NH_2$     $\underset{BOH}{\overset{HX}{\rightleftharpoons}}$     $R-NH_3^+ \ X^-$

$2^0$ Amine     $R-NH-R$     $\underset{BOH}{\overset{HX}{\rightleftharpoons}}$     $R-NH_2-R^+ \ X^-$

$3^0$ Amine     $\underset{R-N-R}{\overset{R}{|}}$     $\underset{BOH}{\overset{HX}{\rightleftharpoons}}$     $\underset{H}{\overset{R}{R-N-R^+}} \ X^-$

$$\underset{R}{\overset{R}{R-N-R^+}} \ X^-$$

Quaternary Ammonium Salt

The quaternary ammonium salts are stable compounds that are not converted to amines by treatment with base. The nitrogen-carbon bonds may be alkyl bonds, aryl bonds, or a mixture of alkyl-aryl bonds. The nomenclature is derived by naming the organic substituents followed by the word ammonium and then the particular salt that is present. An example is the compound tetraethyl ammonium sulfate:

$$\left( \begin{array}{c} C_2H_5 \\ C_2H_5-N-C_2H_5 \\ C_2H_5 \end{array} \right)_2^{++} \qquad SO_4^{=}$$

TEA Sulfate

*B. Physical-Chemical Properties.* While the ammonium salts formed from primary, secondary, and tertiary amines are reversible, as shown, this is not true of quaternary ammonium salts. These salts are relatively stable and require considerable energy to break the carbon-nitrogen bond. The quaternary ammonium salts are ionic compounds that, if capable of dissociation in water, exhibit significant water solubility. Ion-dipole bonding to water of the quaternary ammonium has the potential of dissolving 20 to 30 carbon atoms. Most of the quaternary ammonium salts commonly seen in pharmacy are water soluble.

*C. Metabolism.* There is no special metabolism of quaternary ammonium salts that the student need be familiar with.

**QUESTIONS**

17. Which nitrogen in the compound shown is a tertiary amine?
   1. Nitrogen 1
   2. Nitrogen 2
   3. Nitrogen 3
   4. Nitrogen 4
   5. Nitrogen 5

18. Which nitrogen in the compound is most basic?

    1. Nitrogen 1
    2. Nitrogen 2
    3. Nitrogen 3
    4. Nitrogen 4
    5. Nitrogen 5

19. Which nitrogen in the compound is least basic?

    1. Nitrogen 1
    2. Nitrogen 2
    3. Nitrogen 3
    4. Nitrogen 4
    5. Nitrogen 5

20. What type(s) of metabolism is possible at nitrogen 2?

    1. Deamination
    2. Methylation
    3. Sulfate conjugation
    4. Glucuronic acid conjugation
    5. Stable nitrogen, no metabolism

# 11

# Carboxylic Acids

*A. Nomenclature.* A carboxylic acid is a molecule that contains a characteristic carboxyl group to which may be attached a hydrogen, alkyl, aryl, or heterocyclic system. The common nomenclature of the carboxylic acids is used more often than with most other

-Common

$$\overset{O}{\underset{}{HC}}-OH$$           Formic Acid

$$CH_3-\overset{O}{\underset{}{C}}-OH$$           Acetic Acid (Vinegar)

$$CH_3-CH_2-\overset{O}{\underset{}{C}}-OH$$           Propionic Acid

$$CH_3-CH_2-CH_2-CH_2-CH_2-\overset{O}{\underset{}{C}}-OH$$           Caproic Acid

-IUPAC (Alkan<u>oic</u> Acid)

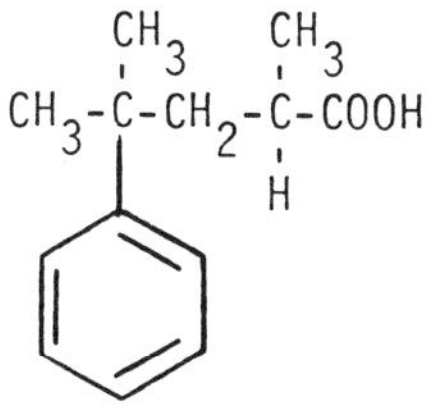

$$CH_3-\overset{CH_3}{\underset{}{C}}-CH_2-\overset{CH_3}{\underset{H}{C}}-COOH$$           2,4-Dimethyl-4-phenylpentan<u>oic</u> Acid

5   4  3   2 1

functional groups, probably because of the wide variety of carboxylic acids found in nature. Even without branching of the alkyl
chain, this nomenclature becomes difficult to remember with such
uncommon names as caproic, caprylic, capric, and lauric acids. The
official nomenclature returns to the use of the hydrocarbon names
such as methane, ethane, propane, butane, and pentane. As with all
IUPAC nomenclature, the longest continuous chain containing the
functional group, in this case the carboxyl group, is chosen as the
base unit. The hydrocarbon name is used, the "e" is dropped and
replaced with "oic," which signifies a carboxyl group, and this is
followed by the word acid. The numbering always starts with the
carboxyl group. This is illustrated on page 57.

   *B. Physical-Chemical Properties.*   The carboxylic acid functional
group consists of a carbonyl and a hydroxyl group; both, when taken
individually, are polar groups that can hydrogen bond. The hydrogen of the -OH can hydrogen bond to either of the oxygen groups
in another carboxyl function (Fig. 11–1). The amount and strength of
hydrogen bonding in the case of a carboxylic acid are greater than in
the case of alcohols or phenols because of the greater acidity of the
carboxylic acid and because of the additional sites of bonding. From
this discussion, it would be predicted that carboxylic acids are
high-boiling liquids and solids. If the carboxyl can strongly hydrogen bond to itself, then it is reasonable to predict that the carboxyl
group can hydrogen bond to water, resulting in water solubility. In
Table 11–1, the effect of the strong intermolecular hydrogen bonding can be seen by examining the boiling points of several of the
carboxylic acids, while the strong hydrogen bonding to water is
demonstrated by the solubility of the carboxylic acids in water. Once
again, as the lipophilic hydrocarbon chain length increases, the
water solubility decreases drastically. A carboxyl group will solubilize at a 1% concentration approximately five carbon atoms.

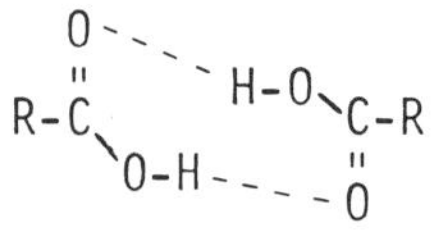

High Boiling Point

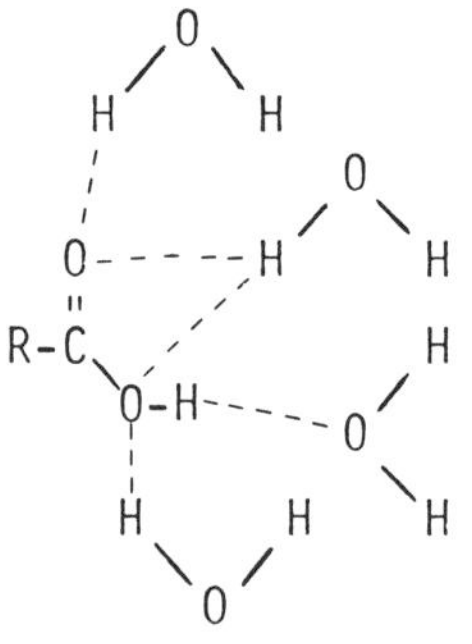

Water Solubility

Fig. 11–1.   Intermolecular bonding of carboxylic acids

**Table 11-1.**
Boiling Points and Water Solubility of Common
Organic Acids

| $\overset{O}{\overset{\|}{R-C-OH}}$ | Boiling Point $^{O}C$ | Solubility | |
|---|---|---|---|
| | | (g/100g $H_2O$) | (g/100g EtOH) |
| H | 100.5 | $\infty$ | $\infty$ |
| $CH_3$ | 118.0 | $\infty$ | $\infty$ |
| $CH_3-CH_2$ | 141.0 | $\infty$ | $\infty$ |
| $CH_3-(CH_2)_2$ | 164.0 | $\infty$ | $\infty$ |
| $CH_3-(CH_2)_3$ | 187.0 | 3.7 | Soluble |
| $CH_3-(CH_2)_4$ | 205.0 | 1.0 | Soluble |
| $C_6H_5$ | 250.0 | 0.34 | Soluble |
| $CH_3-(CH_2)_8$ | | 0.015 | Soluble |
| $CH_3-(CH_2)_{10}$ | | Insoluble | 100 |
| $CH_3-(CH_2)_{12}$ | | Insoluble | Soluble |
| $CH_3-(CH_2)_{16}$ | | Insoluble | 5.0 |

Another solvent important in pharmacy is ethanol. Ethanol has both a hydrophilic and lipophilic portion, and bonding between an organic molecule and ethanol therefore may involve both dipole-dipole bonding and van der Waals bonding. It is not surprising, then, that the solubility of the carboxylic acids is much greater in ethanol than it is in water. Although pure ethanol cannot be used internally, ethanol-water combinations can and greatly increase the solution potential of many drugs.

Turning now to an extremely important property of the carboxylic acids, their acidic property, one sees the familiar dissociation of a carboxylic acid (giving up a proton) shown in Table 11-2. This dissociation, by definition, makes the group an acid.

From general chemistry it will be recalled that the strength of an acid depends on the concentration of protons in solution, which depends on dissociation. The value of $K_1$ and $K_{-1}$ in turn depends on the stability of the carboxylate anion in relation to the undissociated carboxylic acid. In other words, if we are considering two acids, acid 1 (in which the carboxylate anion is unstable) and acid 2 (in which

**Table 11–2.**

Dissociation Constants and $pK_a$ Values in Water of Common
Carboxylic Acids

$$R\text{-}\overset{O}{\overset{\|}{C}}\text{-}OH \quad + \quad H_2O \quad \rightleftharpoons \quad H_3O^+ \quad + \quad R\text{-}\overset{O}{\overset{\|}{C}}\text{-}O^- \quad \longleftarrow \longrightarrow \quad R\text{-}C\underset{O}{\overset{O}{\diagup}} \quad (-)$$

Example 1: $\quad R \rightarrow \overset{O}{\overset{\|}{C}}\text{-}OH \quad + \quad H_2O \rightleftharpoons H_3O^+ \quad + \quad R \rightarrow \overset{O}{\overset{\|}{C}}\text{-}O^-$

Example 2: $\quad R \leftarrow \overset{O}{\overset{\|}{C}}\text{-}OH \quad + \quad H_2O \rightleftharpoons H_3O^+ \quad + \quad R \leftarrow \overset{O}{\overset{\|}{C}}\text{-}O^-$

| $R\text{-}\overset{O}{\overset{\|}{C}}\text{-}OH$ | Dissociation Constant (Ka and pKa) | |
|---|---|---|
| | Ka (In water) | pKa |
| H | $17.7 \times 10^{-5}$ | 3.75 |
| $CH_3$ | $1.75 \times 10^{-5}$ | 4.76 |
| $Cl\text{-}CH_2$ | $1.36 \times 10^{-3}$ | 2.87 |
| $Cl_2CH$ | $5.53 \times 10^{-2}$ | 1.26 |
| $Cl_3C$ | $2.32 \times 10^{-1}$ | 0.64 |
| $C_6H_5$ | $6.3 \times 10^{-5}$ | 4.21 |
| $p\text{-}CH_3\text{-}C_6H_4$ | $4.2 \times 10^{-5}$ | 4.38 |
| $m\text{-}CH_3\text{-}C_6H_4$ | $5.4 \times 10^{-5}$ | 4.27 |
| $p\text{-}O_2N\text{-}C_6H_4$ | $3.6 \times 10^{-4}$ | 3.44 |
| $m\text{-}O_2N\text{-}C_6H_4$ | $3.2 \times 10^{-4}$ | 3.50 |

the carboxylate anion is stable), acid 2 with the more stable carboxylate will dissociate to a greater extent, giving up a higher concentration of protons, and therefore is a stronger acid. It has been found
that the nature of the R-group *does* influence the stability of the
carboxylate anion, and it does so in the following manner: if R is an
electron donor group, as shown in Table 11–2, it will destabilize the
carboxylate anion and thus decrease the acidity (this is represented
by the dissociation arrows). To understand how this comes about,

one must look at the carboxylate anion. This anion is stabilized by resonance with the negative charge not remaining fixed on the oxygen but instead being spread across the oxygen-carbon-oxygen. Now, if one considers the effect of pushing electrons toward a region already high in electron density, repulsion occurs. This is an unfavorable situation. In the nonionic carboxylic acid form, resonance stabilization is not occurring, and the problem is reduced. Therefore, in example 1, the nonionic form is more stable than the ionic form. In example 2, the opposite effect is considered, electron withdrawal by the R-group. If electron density around the carbonyl carbon is reduced, this should increase the ease of resonance stabilization, in turn increasing the stability of the carboxylate anion. If one consid-

**Table 11–3.**
Solubility Properties of Sodium Salts of Common Organic Acids

| $R-C-O^- Na^+$ | Solubility (g/100g $H_2O$) |
|---|---|
| $C_6H_5$ | 55.5 |
| $CH_3$ | 125.0 |
| $CH_3-CH_2$ | 100.0 |
| $CH_3-(CH_2)_{16}$ | 10.0 |

ers example 2 in relationship to example 1, acid 2 would be predicted to be more acidic than acid 1. Table 11–2 has examples of compounds that fit this description. The methyl group is an electron donor that reduces the acidity with respect to that of formic acid, while the phenyl can be considered an electron sink or, with respect to alkyl acids, an electron-withdrawing group; therefore, benzoic acid is a stronger acid than acetic acid. The addition of halogens to an alkyl changes the nature of the alkyl. In chloroacetic acid, the chloride, being electronegative, pulls electrons away from the carbon, which in turn pulls electrons away from the carbonyl. This effect is quite strong, as is seen in the $K_a$. This electron-withdrawing effect continues to increase, as the number of halogens increases, to give a strong carboxylic acid, trichloroacetic acid.

As discussed earlier for phenols and aromatic amines, substitution on the aromatic ring of benzoic acid will influence acidity. Ortho substitution is not always predictable, but in most cases the acidity of the acid is increased by ortho substitution. Meta and para substitution is predictable. Substitution on the benzene ring with an electron-releasing group decreases acidity. If this substituent is para, the decrease in acidity with respect to benzoic acid will be greater than if the substituent is meta. If the substituent is an electron-withdrawing group, the acidity of the acid will increase. The

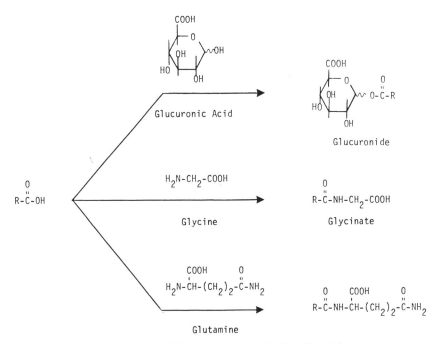

Fig. 11–2.  Metabolic conjugation of carboxylic acids

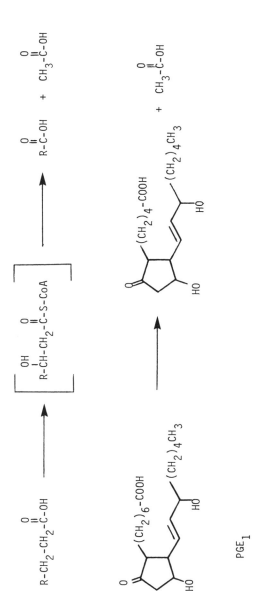

Fig. 11–3.   Beta oxidation of alkyl carboxylic acids

greatest increase is observed when the substituent is para. One should recall that this is the same trend seen for substituted phenols.

One additional property of carboxylic acids is their reactivity toward base. Carboxylic acids will react with a base to give a salt, as shown in Table 11–3. If one is considering water solubility, the interaction of a salt with water through dipole-ion bonding is much stronger than dipole-dipole interaction of the acid. Therefore, a considerable increase in water solubility should and does occur. The same point must be made here as was made with phenol and amine salts: the salt must be able to dissociate in order to dissolve in water. Salts formed from carboxylic acids and sodium, potassium, or ammonium hydroxide show a great increase in water solubility. Salts formed with heavy metals tend to be relatively insoluble. Examples of such insoluble salts are the heavy metal salts (e.g., calcium, magnesium, zinc, aluminum) of carboxylic acids. When salts of carboxylic acids dissolve in water, a characteristic alkaline pH is common. With sodium and potassium salts, the pH is generally quite high. As with other salts, if acid is now added to this solution, one can reverse the carboxylic acid-base reaction and regenerate the carboxylic acid. The free acid is less soluble than was the salt, and precipitation may result. This is an important chemical incompatibility that one should keep in mind when dealing with water-soluble carboxylate salts. In summary, it is found that amines and carboxylic acids are common functional groups found in drugs. These groups have a potentiating effect on solubility, and both groups can form salts that, if capable of dissociation, will greatly increase water solubility.

*C. Metabolism.*     The metabolism of the carboxylic acid is relatively simple. Carboxylic acids can undergo a variety of conjugation reactions. They can conjugate with glucuronic acid to form glucuronides and also with amino acids (Fig. 11–2). Glycine and glutamine are two common amino acids that form conjugates with acids.

Another common type of metabolism of alkyl carboxylic acids is oxidation beta to the carboxyl group. This is a common reaction in the metabolism of fatty acids. The reaction proceeds through a sequence in which the carboxyl group is bound to coenzyme A (CoA). The bound acid is oxidized to enoyl CoA, hydrated to $\beta$-hydroxyacetyl CoA, oxidized to $\beta$-ketoacyl CoA, and finally cleaved to the shortened carboxylic acid plus acetic acid, as shown in Figure 11–3.

**QUESTIONS**

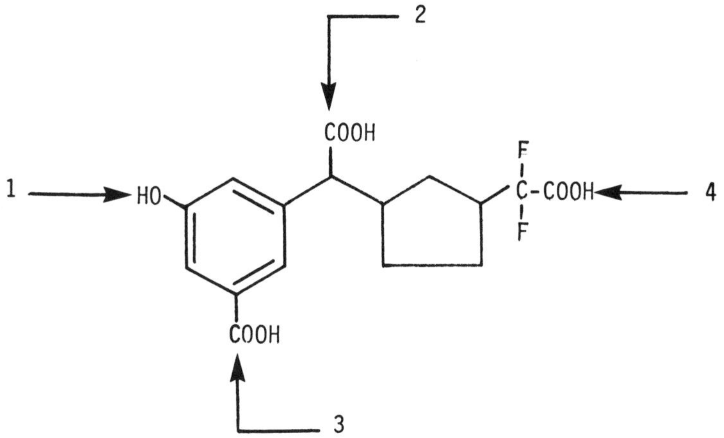

21. Which acid in the compound shown is most acidic?

    1. Acid 1
    2. Acid 2
    3. Acid 3
    4. Acid 4

22. Which acid is least acidic?

    1. Acid 1
    2. Acid 2
    3. Acid 3
    4. Acid 4

23. You would predict that the compound would be soluble in:

    1. Water
    2. Aqueous sodium hydroxide
    3. Aqueous hydrochloric acid
    4. No aqueous media

24. What type of metabolism(s) is expected for functional group 2?

    1. Sulfate conjugation
    2. Glucuronide conjugation
    3. Methylation
    4. Hydrolysis
    5. Glutamine conjugation

# 12

# Functional Derivatives of Carboxylic Acids

In discussing the carboxylic acids, it is necessary to discuss several derivatives of carboxylic acids. The first to be discussed will be the esters.

$$CH_3-CH_2-\overset{\overset{\displaystyle O}{\|}}{C}-O-CH\overset{\diagup CH_3}{\diagdown CH_3}$$

$$CH_3-\overset{\overset{\displaystyle O}{\|}}{C}-O-\overset{\overset{\displaystyle CH_3}{|}}{\underset{\underset{\displaystyle CH_3}{|}}{C}}-CH_3$$

Common:  Isopropyl propion<u>ate</u>

IUPAC :  2-Propyl propano<u>ate</u>

t-Butyl acet<u>ate</u>

2-Methyl-2-propyl ethano<u>ate</u>

## 1. Esters

*A. Nomenclature.*  The nomenclature consists of combining the alcohol and carboxylic acid nomenclature as either common nomenclature or as IUPAC nomenclature, but not mixed. The name of the alcohol radical comes first, followed by a space, and then the name of the acid. To show that it is a functional derivative of an acid, the "ic" ending of the acid is dropped and replaced by the "ate." If you do not remember alcohol and acid nomenclature, you should return to the appropriate sections and review this material.

*B. Physical-Chemical Properties.*  The physical properties of the esters are rather interesting and show a similarity to the ethers. In formation of esters, a polar alcohol is combined with a polar acid to give a much less polar, low-boiling liquid. In the case of ethers, two alcohols were joined with the same decrease in polarity and boiling points. As with ethers, in the ester, the two hydroxyl groups necessary for intermolecular hydrogen bonding are destroyed, and along with this goes the loss of the intermolecular bonding and a decrease in water solubility. Proof of this effect can be seen in Table 12–1. The

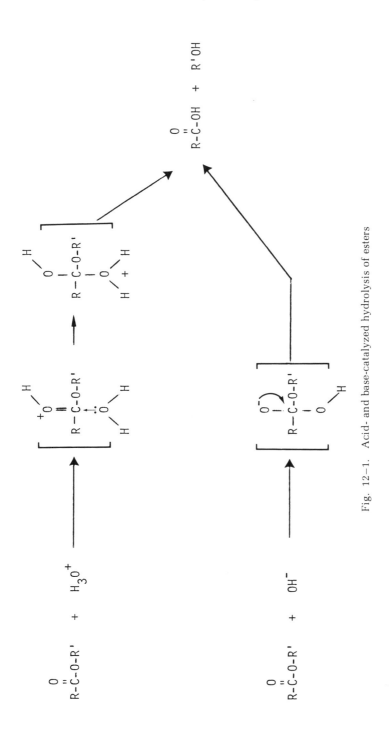

Fig. 12–1. Acid- and base-catalyzed hydrolysis of esters

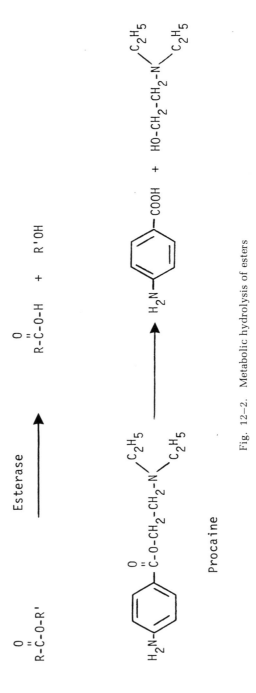

Fig. 12–2.   Metabolic hydrolysis of esters

**Table 12–1.**
Boiling Points and Water Solubility of Common Esters

| R | R' | Boiling Point $^{o}$C | Solubility (g/100g $H_2O$) |
|---|---|---|---|
| $CH_3$ | $CH_3$ | 57.5 | ∞ |
| $CH_3$ | $CH_3-CH_2$ | 77.0 | 10.0 |
| $CH_3-CH_2$ | $CH_3$ | 79.7 | 6.25 |
| $CH_3$ | $CH_3-CH_2-CH_2$ | 102.0 | 1.60 |
| $CH_3$ | $CH_3-CH_2-CH_2-CH_2$ | 126.0 | 0.83 |
| $CH_3-(CH_2)_3$ | $CH_3$ | 102.0 | 1.67 |
| $C_6H_5$ | $CH_3$ | 198.0 | Insoluble |

(Ester structure: R-C(=O)-O-R')

boiling point of acetic acid is 118°, which itself is above that of many of the esters. The water solubility of esters is due to hydrogen bonding between the hydrogen of water and the electron-dense oxygen of the ester carbonyl. While esters are not highly water soluble, they are quite soluble in alcohol.

An important chemical property that most esters display is the ease of hydrolysis back to the alcohol and the carboxylic acid. Esters are especially prone to base-catalyzed hydrolysis but will also hydrolyze in the presence of acid and water (Fig. 12–1). What this means to medicinal chemistry is that esters are unstable in the presence of basic media in vitro and must therefore be protected from strongly alkaline conditions.

C. *Metabolism.* Hydrolyzing enzymes in the body carry out hydrolysis through a base-catalyzed mechanism. It is therefore not unexpected that esters are unstable in the body and are converted to the free acid and the alcohol. In many cases, carboxylic acids are synthetically prepared and administered as esters, even though the active drug is the acid. It is known that the acid will be regenerated metabolically.

### 2. Amides

A. *Nomenclature.* The second important functional derivative of the carboxylic acid is the amide group shown in the following illustration. An example of the common and official nomenclature is

$$CH_3-CH_2-CH_2-CH_2-\overset{\overset{O}{\|}}{C}-N\underset{CH_3}{\overset{\overset{\overset{CH_3}{|}}{CH-CH_3}}{<}}$$

Common: N-Methyl-N-isopropyl valeramide

IUPAC  : N-Methyl-N-2-propyl pentanamide

$$\text{(phenyl)}-\overset{\overset{O}{\|}}{C}-N\underset{\text{(phenyl)}}{\overset{CH_3}{<}}$$

N-Methyl-N-phenyl benzamide

N-Methyl-N-phenyl benzamide

also shown in the illustration. In the case of the common nomenclature, the common name of the amine followed by the common name of the acid is used. The ending "ic" of the acid is then dropped and replaced by the word "amide." The same approach is used for official nomenclature except that the official name of the amine and the official name of the acid are used. The "oic" ending is dropped and replaced by "amide."

B. *Physical-Chemical Properties and Metabolism.* The physical properties of the amides are much different than might have been

**Table 12–2.**
Boiling Points and Water Solubility of Common Amides

$$R_1-\overset{\overset{O}{\|}}{C}-N\underset{R_3}{\overset{R_2}{<}}$$

| $R_1$ | $R_2$ | $R_3$ | Boiling Point °C | Solubility (g/100g $H_2O$) |
|-------|-------|-------|------------------|----------------------------|
| H | H | H | 210 | Soluble |
| H | $CH_3$ | H | 180 | Soluble |
| H | $CH_3$ | $CH_3$ | 153 | Soluble |
| $CH_3$ | H | H | 222 | 200 |
| $CH_3$ | $CH_3$ | H | 210 | Soluble |
| $CH_3$ | $CH_3$ | $CH_3$ | 163 | Soluble |
| $C_6H_5$ | H | H | 288 | 1.35 |
| H | $C_6H_5$ | H | 271 | 2.86 |
| $CH_3$ | $C_6H_5$ | H | 304 | 0.53 |

predicted after the earlier discussions of esters. Like the esters, the amides have a polar carboxylic acid combined with the weakly polar primary or secondary amine or ammonia to give the mono- substituted, disubstituted, or unsubstituted amides, respectively. The resulting amides still possess considerable polarity, as indicated by the high boiling points and water solubility (Table 12–2). These properties are quite different from those of esters. It is interesting to note that, with any series, as the substitution on the nitrogen in- creases, the boiling point decreases. As an example, look at form- amide, N-methyl formamide, and N,N-dimethylformamide. This may be explained in part by a consideration of the resonance forms of amides, as shown in Figure 12–3. The unshared pair of electrons of the nitrogen no longer remain on the nitrogen but are spread across the nitrogen, carbon, and oxygen. This has a significant effect on the polarity of the amide. Since boiling points depend on the amount and strength of intermolecular bonding, the unsubstituted and monosubstituted amide would be expected to show strong intermolecular bonding owing to the high electron density on oxy- gen bonding to the hydrogens or hydrogen on the nitrogen. In the case of the disubstituted amides, both hydrogens have been replaced on the nitrogen, and intermolecular hydrogen bonding is not possi- ble. Disubstituted amides are still capable of dipole-dipole bonding,

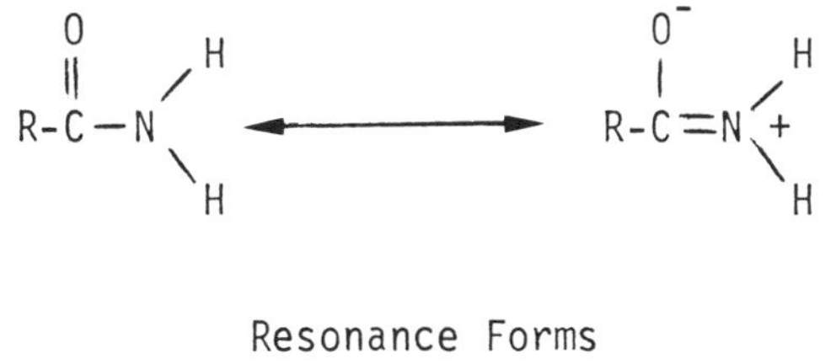

Resonance Forms

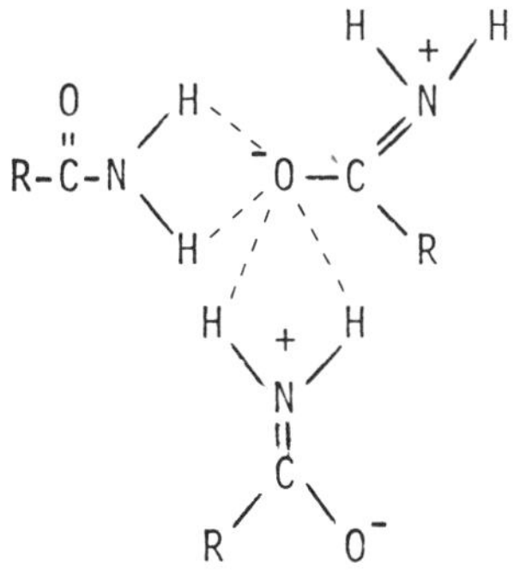

Fig. 12–3.   The effect of resonance structures of amides on intermolecular hydrogen bonding

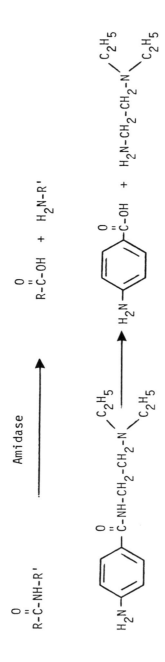

Fig. 12—4.   Metabolic hydrolysis of amides

but not of hydrogen bonding. Thus, the boiling point drops. Water solubility requires only a polar material, since the hydrogen can be supplied by the water. Both substituted and unsubstituted amides can hydrogen bond to water through the hydrogen of water and show good water solubility. As the hydrocarbon portion of the amide increases, so the lipophilic nature increases and water solubility decreases.

A chemical property that differentiates amides from esters is the greater stability of the amide. Amides are relatively stable to the acid, base, and enzymatic conditions encountered in pharmacy. The reason for this stability again can be related to the resonance forms of the amide with its overlapping clouds of electrons. The importance of the increase in stability of the amide over the ester has been used to advantage in preparing drugs with prolonged activity. Although amides are relatively stable to acid, base, and enzymes, a metabolism that may be encountered is hydrolysis catalyzed by the amidase enzymes. An example is illustrated in Figure 12–4. The important point to remember is that amides are more stable in vivo than are esters.

One additional point concerning the amides is that, although basic amines were used to prepare the amides, amides are nearly neutral functional groups, and therefore acid salts cannot be formed. Returning to the definition of a base, an unshared pair of electrons is essential for basicity. The unshared pair of electrons must be available for donation, a situation that does not exist in amides. In the amide, the pair of electrons no longer remain on the nitrogen but are spread over the nitrogen, carbon, and the oxygen. This resonance of the electrons reduces their availability and thus the amide's basicity.

### 3. Carbonates, Carbamates, and Ureas

The final functional derivatives of the carboxylic acids will be grouped together because of their similarity to the previously discussed esters and amides. The carbonates, carbamates, and ureas are shown below.

*A. Nomenclature.* Carbonate nomenclature may be common or official. In either case, the two alcohol portions are named and com-

$$\overset{\displaystyle O}{\underset{\displaystyle \quad}{R\text{-}O\text{-}\overset{\|}{C}\text{-}O\text{-}R'}}$$

Carbonate

$$CH_3\text{-}\overset{\overset{\displaystyle CH_3}{|}}{CH}\text{-}O\text{-}\overset{\overset{\displaystyle O}{\|}}{C}\text{-}O\text{-}CH_2\text{-}CH_3$$

Common:  Ethyl isopropyl <u>carbonate</u>

IUPAC :  Ethyl 2-propyl <u>carbonate</u>

bined with the word carbonate (carbonates are ester derivatives of carbonic acid). The example shows both the common and official names.

Carbamates are ester-amide derivatives of carbonic acid and, like carbonates, require the naming of the alcohol and the amine, either using common names or IUPAC nomenclature, followed by the word carbamate.

The ureas, the diamide derivatives of carbonic acid, are illustrated as follows. In this case, it may be necessary, as we see in the

$$R-O-\overset{\overset{\displaystyle O}{\|}}{C}-N\overset{\displaystyle R'}{\underset{\displaystyle R''}{<}}$$

Carbamate

$$CH_3-CH_2-O-\overset{\overset{\displaystyle O}{\|}}{C}-N\overset{\displaystyle CH_3}{\underset{\displaystyle \underset{CH_3}{\overset{|}{C}-CH_3}}{<}}$$

Common:  Ethyl N-methyl-N-t-butyl <u>carbamate</u>

IUPAC :  Ethyl N-methyl-N-2-(2-methylpropyl) <u>carbamate</u>

example, to differentiate between the two nitrogens. This is done by using N for the one nitrogen and N′ for the other nitrogen. In the preceding cases, the structures should be obvious by the use of the terms carbonate, carbamate, and urea in the nomenclature.

$$\overset{\displaystyle R}{\underset{\displaystyle R}{>}}N-\overset{\overset{\displaystyle O}{\|}}{C}-N\overset{\displaystyle R'}{\underset{\displaystyle R'}{<}}$$

Urea  (R = R′ = H)

$$\overset{\displaystyle CH_3}{\underset{\displaystyle C_2H_5}{>}}N-\overset{\overset{\displaystyle O}{\|}}{C}-N\overset{\displaystyle CH_3}{\underset{\displaystyle C_2H_5}{<}}$$

N,N′-Dimethyl-N,N′-diethyl <u>urea</u>

*B. Physical-Chemical Properties.*  The physical and chemical properties of the carbonate parallel those of the ester, while the properties of the urea are similar to those of the amide. The carbamate has physical properties that represent the combined effect of both components. Chemically, however, the carbamate shares reactions more like those of an ester, by which is meant that carbonates

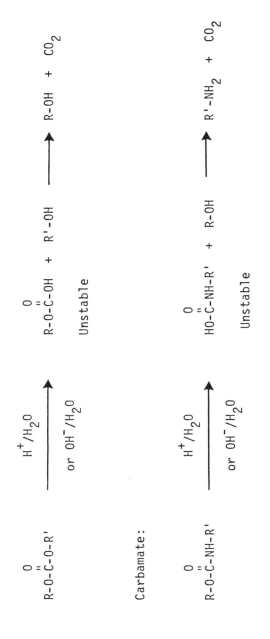

Carbonate:

$$R-O-\overset{\overset{\text{O}}{\|}}{C}-O-R' \quad \xrightarrow[\text{or } OH^-/H_2O]{H^+/H_2O} \quad R-O-\overset{\overset{\text{O}}{\|}}{C}-OH + R'-OH \quad \longrightarrow \quad R-OH + CO_2$$

Unstable

Carbamate:

$$R-O-\overset{\overset{\text{O}}{\|}}{C}-NH-R' \quad \xrightarrow[\text{or } OH^-/H_2O]{H^+/H_2O} \quad HO-\overset{\overset{\text{O}}{\|}}{C}-NH-R' + R-OH \quad \longrightarrow \quad R'-NH_2 + CO_2$$

Unstable

Ureas : Relatively inert

Fig. 12–5.   Acid- or base-catalyzed hydrolysis of carbonates and carbamates

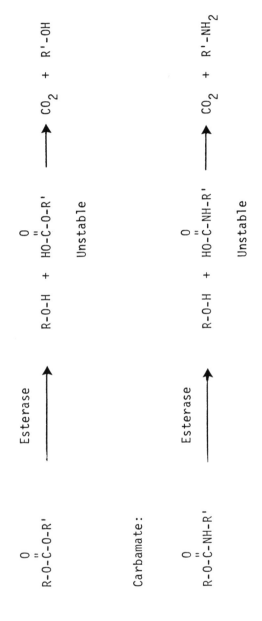

Carbonates:

$$R-O-\overset{\overset{\displaystyle O}{\|}}{C}-O-R'$$

$\xrightarrow{\text{Esterase}}$

$R-O-H \quad + \quad HO-\overset{\overset{\displaystyle O}{\|}}{C}-O-R'$

Unstable

$\xrightarrow{\hspace{2cm}} CO_2 \quad + \quad R'-OH$

Carbamate:

$$R-O-\overset{\overset{\displaystyle O}{\|}}{C}-NH-R'$$

$\xrightarrow{\text{Esterase}}$

$R-O-H \quad + \quad HO-\overset{\overset{\displaystyle O}{\|}}{C}-NH-R'$

Unstable

$\xrightarrow{\hspace{2cm}} CO_2 \quad + \quad R'-NH_2$

Ureas : Relatively stable

Fig. 12–6.  Metabolic hydrolysis of carbonates and carbamates

and carbamates are unstable to acid and base conditions. The ureas, being similar to amides, are relatively nonreactive solids with the polarity properties previously discussed for amides (Fig. 12–5).

*C. Metabolism.* The metabolic route open to these functional derivatives of carbonic acid is the hydrolysis reaction. This reaction is catalyzed by the esterase enzymes. Both the carbonate and the carbamate have the ester portion first hydrolyzed to give the monosubstituted carbonic acid. This acid is unstable and decomposes with loss of $CO_2$, as shown in Figure 12–6. Carbonates are hydrolyzed with formation of carbon dioxide and alcohol, and carbamates decompose to carbon dioxide, an alcohol, and an amine. The ureas are relatively stable chemicals and are not commonly metabolized by hydrolysis.

**QUESTIONS**

$$\begin{array}{c} O \\ \parallel \\ C\text{-}O\text{-}CH_2CH_2CH_3 \end{array}$$

$$\begin{array}{c} C\text{-}NH\text{-}CH_2CH_2CH_3 \\ \parallel \\ O \end{array}$$

25. Predict the solubility of the compound shown.
   1. Soluble in water
   2. Soluble in aqueous hydrochloric acid
   3. Soluble in aqueous sodium hydroxide
   4. Soluble in all of the above
   5. Insoluble in all of the above

26. Predict the in vitro (shelf) instability of the above compound.
   1. Unstable in aqueous acid due to ester hydrolysis
   2. Unstable in aqueous acid due to amide hydrolysis
   3. Unstable in aqueous base due to ester hydrolysis
   4. Unstable in aqueous base due to amide hydrolysis

27. Predict the metabolism of the ester.
   1. Hydrolyzed by esterase
   2. Prone to oxidation
   3. Undergoes reduction
   4. Peroxide formation

28. Predict the metabolism of the amide.
   1. Dealkylation
   2. Glucuronide formation
   3. Hydrolyzed by amidase
   4. Sulfate conjugation

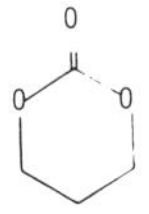

I. 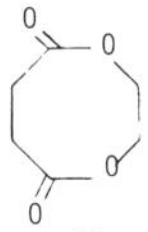    II. 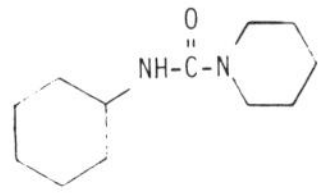    III. 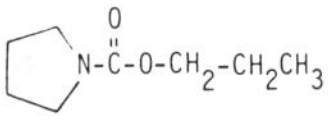    IV.

29. Which of the compounds shown contains a carbamate?

   1. Compound I
   2. Compound II
   3. Compound III
   4. Compound IV

30. What metabolism is expected for compound III?

   1. Hydrolysis by esterase
   2. Hydrolysis by amidase
   3. N-dealkylation
   4. Conjugation with glucuronic acid, sulfuric acid, etc.
   5. Fairly stable

31. What metabolism is expected for compound I?

   1. Hydrolysis by esterase
   2. Hydrolysis by amidase
   3. Peroxide formation
   4. Conjugation with glucuronic or sulfuric acid
   5. Stable

# 13

## Sulfonic Acids and Sulfonamides

Benzene sulfonic acid     Methane sulfonic acid     p-Toluene sulfonic acid

*A. Nomenclature.* The nomenclature for the sulfonic acids is quite simple and is illustrated for benzenesulfonic acid, methanesulfonic acid, and p-toluenesulfonic acids, three of the most common sulfonic acids.

*B. Physical-Chemical Properties.* Concerning the physical and chemical properties, it is found that these three acids are strong acids, nearly as strong as the mineral acids. Saying that an acid is a strong acid implies that considerable dissociation occurs, and this suggests the possibility of ion-dipole interaction with water. By now we know that such binding will favor water solubility. The solubility of several of these acids can be seen from the data presented in Table 13–1.

Common: Benzene sulfonamide       p-Toluene sulfonamide

IUPAC : Benzene sulfonamide       4-Methylbenzene sulfonamide

Common: p-Aminobenzene sulfonamide

Sulfanilamide

IUPAC : 4-Aminobenzene sulfonamide

With this brief background, let us look at a highly important derivative of the sulfonic acids, the sulfonamides. This group of com-

**Table 13–1.**
Water Solubility of Common Sulfonic Acids

| $\begin{array}{c} O \\ \| \\ R\text{-}S\text{-}OH \\ \| \\ O \end{array}$ | Solubility (g/100g $H_2O$) | Ka in $H_2O$ |
|---|---|---|
| $CH_3$ | 20.0 | |
| $C_6H_5$ | Soluble | $2 \times 10^{-1}$ |
| $p\text{-}CH_3\text{-}C_6H_4$ | 67.0 | |
| $C_{10}H_7$ (naphthyl) | Soluble | |

pounds is important in medicinal chemistry, since a wide variety of drugs have the benzenesulfonamide nucleus. The nomenclature is fairly straightforward. The name of the substituted benzene followed by the word sulfonamide is commonly used, although a few important common names will also have to be memorized, such as sulfanilamide. The properties of the sulfonamides are distinctive. The benzenesulfonamides tend to be solids with high melting points and poor water solubility. A property that they share in common with amides is their chemical stability. Benzenesulfonamides are quite stable to acid, base, or enzymatic hydrolysis.

Water soluble

Fig. 13–1.   Salt formation of benzenesulfonamides

A highly significant chemical property of aryl sulfonamides is the ability to form salts. Salt formation occurs with bases but not acids. The aryl sulfonamides are weak acids and react with strong bases, such as sodium or potassium hydroxide, to give sodium or potassium salts. These salts are highly water soluble. As with salts of carboxylic acids, these salts, when dissolved in water, will produce an alkaline pH. This is again important in that salts of aryl sulfonamides, like salts of carboxylic acids, are incompatible with acid or acidic systems. In acid media, an acid-base reaction occurs and will give the free sulfonamide and sodium chloride. This is the reverse of the reaction shown in Figure 13–1.

# 14
# Thio Ethers and the Nitro Group

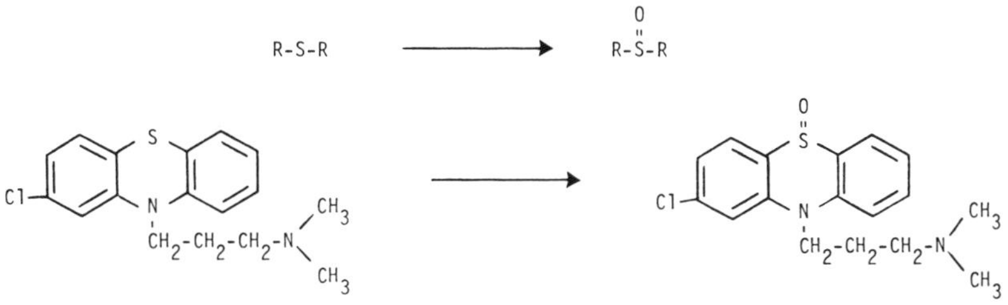

$$R\text{-}S\text{-}R \quad \longrightarrow \quad \overset{\displaystyle \overset{O}{\|}}{R\text{-}S\text{-}R}$$

Chlorpromazine

Chlorpromazine Sulfoxide

Fig. 14–1.  Metabolic oxidation of thioethers

The last two functional groups to be considered are the thio ether and the nitro group, shown in the following two illustrations.

The thio ether is mentioned because it is an important functional group found in several different classes of drugs. The reader should recognize this functional group. An important fact that you should be aware of with this functional group is its metabolism. Thio ethers are commonly oxidized to a sulfoxide or sulfone. An example is given in Figure 14–1.

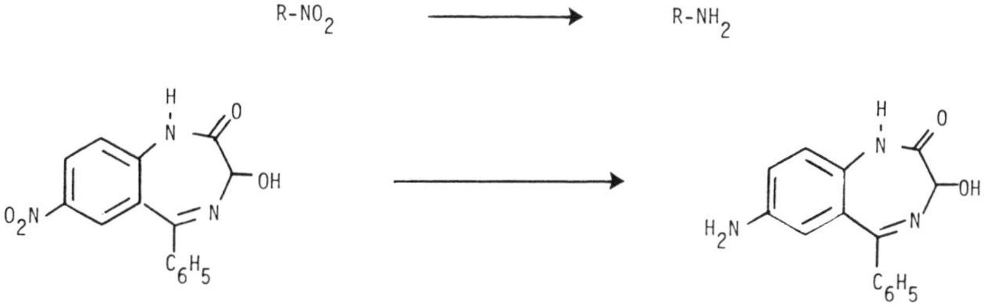

$$R\text{-}NO_2 \quad \longrightarrow \quad R\text{-}NH_2$$

Nitrazepam

Fig. 14–2.  Metabolic reduction of the aromatic nitro group

The other functional group that you should be able to recognize is the nitro group. In most cases the nitro group appears in drugs as the aromatic nitro group. It also has a common metabolism, reduction to the amine, as is shown in Figure 14–2.

## QUESTIONS

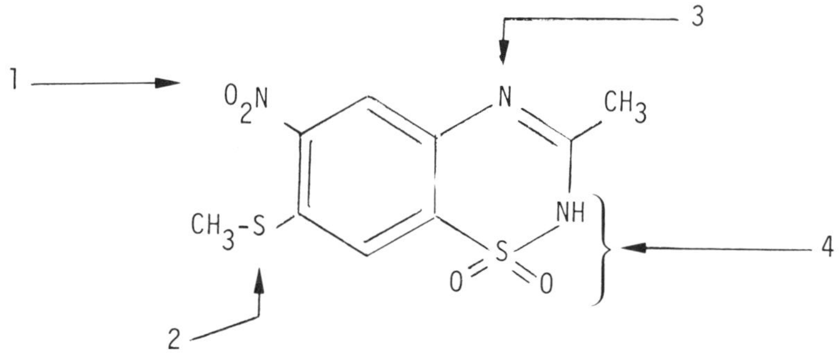

32. The compound shown is soluble in aqueous sodium hydroxide because of salt formation at functional group ____.

    1. Group 1
    2. Group 2
    3. Group 3
    4. Group 4
    5. Not soluble

33. The compound is soluble in aqueous hydrochloric acid because of salt formation at functional group ____.

    1. Group 1
    2. Group 2
    3. Group 3
    4. Group 4
    5. Not soluble

34. Predict the metabolism at position 1.

    1. Oxidation
    2. Reduction
    3. Deamination
    4. Conjugation
    5. Stable

35. Predict the metabolism at position 2.

    1. Oxidation
    2. Reduction
    3. Demethylation
    4. Conjugation
    5. Stable

# 15

## Heterocycles

This chapter introduces the subject of heterocyclic chemistry. Heterocycles are defined as cyclic molecules that contain one or more heteroatoms in a ring. A heteroatom is an atom other than carbon. One need merely glance through an index of biologically active structures to recognize the array of heterocycles found in synthetic and naturally occurring molecules. A background in heterocyclic chemistry is therefore highly desirable. The competencies expected of you from this section of the book consist of:

1. The ability to match a structure of a heterocycle to its common or official name;
2. The ability to list the physical and chemical properties of representative heterocycles;
3. The ability to draw the structure of expected metabolites of common heterocycles.

It would be impossible to introduce all of the possible heterocycles that are of medicinal value within the limitations of this book. I have selected a limited number of monocyclic, bicyclic, and tricyclic rings and will confine the discussion to the heteroatoms of oxygen, nitrogen, and sulfur. In addition, only three-, five-, six-, seven-, and eight-membered monocyclic heterocycles will be considered along with the five-six, six-six, and six-seven bicyclic heterocycles. Several important tricyclic rings will also be considered. In systematic nomenclature, one will recognize a consistent form of nomenclature that follows certain rules for naming the heteroatom and ring size. Tables 15–1 and 15–2 list the rules for heterocyclic systems. In addition, as will be noted in the nomenclature section for the individual heterocycles, the numbering of heterocycles will usually begin with the heteroatom being designated as the 1 position. In heterocycles that contain multiple heteroatoms, the convention is

**Table 15–1.**
Acceptable Prefixes for Common Heteroatoms

| Element | Prefix |
|---------|--------|
| Oxygen | Oxa |
| Nitrogen | Aza |
| Sulfur | Thia |

that, in numbering, the oxygen atom has priority over the sulfur atom, which in turn has priority over the nitrogen atom.

## THREE-MEMBERED RING HETEROCYCLES

### 1. Oxygen

*A. Nomenclature.* A saturated three-membered ring containing oxygen is known as the OXIRANE ring according to the rules presented in Tables 15–1 and 15–2. While correctly named as oxiranes, the common practice for such molecules is to refer to these agents as *epoxides.* A number of natural products can be found that contain the epoxide functional group, and you will be introduced to such compounds in medicinal chemistry.

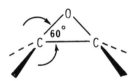

*B. Physical-Chemical Properties.* Epoxides are ethers, but because of the three-membered ring, epoxides have unusual properties. The three-membered ring forces the atoms making up the ring

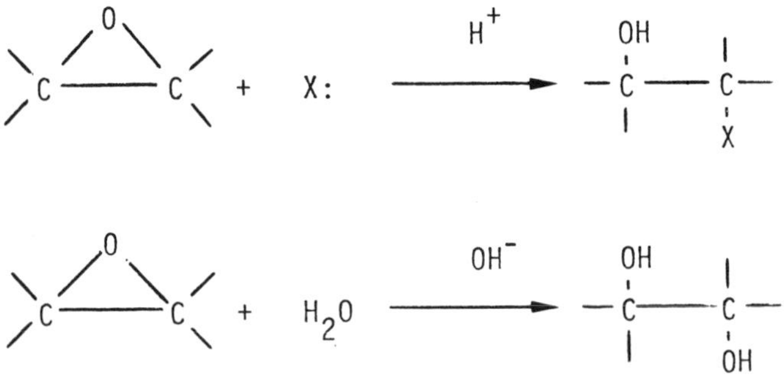

Fig. 15–1. Acid- or base-catalyzed ring opening reactions of epoxides

**Table 15-2.**

Common Suffixes for Nitrogen-Containing Heterocycles and Non-Nitrogen-Containing Heterocycles Based on Ring Size

| Ring Size | Rings With Nitrogen | | | Rings Without Nitrogen | | |
|---|---|---|---|---|---|---|
| | Saturated | Partly Saturated | Unsaturated | Saturated | Partly Saturated | Unsaturated |
| 3 | -iridine | | -irine | -irane | | -irene |
| 5 | -olidine | -oline | -ole | -olane | -olene | -ole |
| 6 | -ine | (di or tetrahydro) | -ine | -ane | (di or tetrahydro) | -ine |
| 7 | (hexahydro) | (di or tetrahydro) | -epine | -epane | (di or tetrahydro) | -epine |
| 8 | (octahydro) | (di, tetra or hexahydro) | -ocine | -ocane | (di, tetra, or hexahydro) | -ocin |

to have an average bond angle of 60°, considerably less than the normal tetrahedral bond angle of 109.5°. This highly strained ring therefore readily opens in the presence of either acid or base catalysts, as shown in Figure 15–1. These reactions are important, since drugs that contain an epoxide ring are also quite reactive both in vitro and in vivo. Such drugs will react with a nucleophile (N:) in the presence of acid or will react with a base that acts as a nucleophile to give open-chain compounds. When drugs containing epoxides are administered to a patient, the epoxide can be expected to react with biopolymers (e.g., proteins), leading to destructive effects on the cell. Such drugs may find use in cancer chemotherapy, but are usually found to be quite toxic.

### 2. Nitrogen

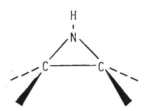

*A. Nomenclature.*    A saturated three-membered ring containing nitrogen is known as the AZIRIDINE ring. This is the only nomenclature used for such a unit. Although aziridine rings are not common in nature nor in many drugs, their intermediacy is required and accounts for the biologic activity of a specific class of anticancer drugs called the nitrogen mustards.

*B. Physical-Chemical Properties.*    Similar to the properties of the epoxide, aziridines are highly strained, highly reactive molecules. The anticancer drug mechlorethamine (Fig. 15–2) is a drug that

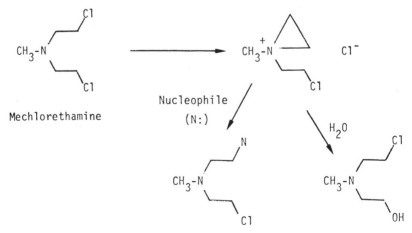

Fig. 15–2.  Aziridine as a reactive intermediate formed from mechlorethamine

owes its activity to the formation of a charged intermediate aziridine (aziridinium ion). Because of its high reactivity, aziridine will react with most nucleophiles (N:), including water. If the nucleophile is part of a biopolymer, this reaction, known as an alkylation reaction, can result in the death of the cell. This reaction is either beneficial, where the cell is a cancer cell, or the mechanism of toxicity where the cell is a host cell. The drug mechlorethamine has only a short half-life when dissolved in water because the aziridinium ion, when formed, reacts with water to give an alcohol that does not possess biologic activity.

## FIVE-MEMBERED RING HETEROCYCLES

### 1. Oxygen

*A. Nomenclature.* Two common five-membered ring hetero-cycles are FURAN and TETRAHYDROFURAN. With this ring system, you need not be concerned with official nomenclature. The common or trivial name will be used in nearly all cases. Substituted furans and tetrahydrofurans are numbered starting with the oxygen as the 1 position and numbering the ring such that any substituents receive the next lowest number.

| IUPAC : | Oxole | Oxolane |
| Common: | Furan | Tetrahydrofuran(THF) |

*B. Physical-Chemical Properties.* Although furan looks like an ether, it does not behave like one. Its properties are more like those of benzene. Furan is an aromatic ring. You may recall that aromatic rings are flat molecules that contain $4N + 2\pi$ electrons, in which $N = 1, 2, 3$, etc. If one considers the $\pi$ electrons or $sp^2$ electrons present in furan, it will be noted that furan contains four $\pi$ electrons in the two double bonds and two pairs of $sp^2$ electrons on the oxygen. One pair of the electrons on oxygen is in the same plane with the four $\pi$ electrons of the two double bonds, thus resulting in a cloud of six electrons located above and below the plane of the ring (Fig. 15–3). Furan therefore has the properties of an aromatic compound, namely, it is relatively nonreactive under the conditions encountered in pharmacy.

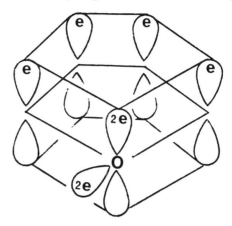

Fig. 15–3.   π electron structure of furan

On the other hand, tetrahydrofuran (THF) has quite different properties when compared to furan. THF is simply a cyclic ether, but unlike its closest open-chain relative, diethylether, THF does not show partial water solubility; it is highly soluble in water. THF is easily oxidized in the presence of air to give peroxides and, like most ethers, also must be protected from atmospheric oxygen.

C. *Metabolism.*   The metabolism of furan and THF follow the pattern expected for an aromatic compound and an ether, respectively. While THF is relatively stable in vivo, furan may undergo the expected aromatic hydroxylation (Fig. 15–4).

Fig. 15–4.   Metabolic hydroxylation of furan

## 2. Nitrogen

A. *Nomenclature.*   Two common five-membered ring heterocycles containing nitrogen are PYRROLE and PYRROLIDINE. Here again, the common name should be learned. The official name can be neglected since it is seldom used. For substituted pyrroles or pyrrolidines, the numbering starts with nitrogen and proceeds clockwise or counterclockwise to give any substituent present the next lowest number.

B. *Physical-Chemical Properties.*   Pyrrole, like furan, is an aromatic compound. It is an aromatic compound that is a weak base and, for our purposes, will be considered neutral. This property can be explained by accounting for all of the nonbonding electrons pres-

|         |            |              |
|---------|------------|--------------|
| IUPAC : | Azole      | Azolidine    |
| Common: | Pyrrole    | Pyrrolidine  |

ent in pyrrole (Fig. 15–5). Nitrogen's extra pair of electrons, which are usually available for sharing and account for the basic properties of amines, are not available for sharing. This pair of electrons is part of the $\pi$ cloud of electrons.

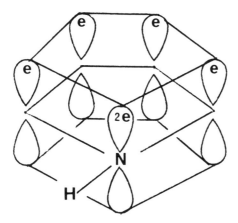

Fig. 15–5.  $\pi$ electron structure of pyrrole

In the fully reduced heterocycle, pyrrolidine, one is dealing with a secondary amine with properties equivalent to any other secondary amine. Unlike pyrrole, with a $K_b$ of approximately $10^{-14}$, pyrrolidine is a strong base, with a $K_b$ of approximately $10^{-3}$. As might be expected, pyrrolidine, with only four carbon atoms and the availability of an unshared pair of electrons for hydrogen bonding, is quite water soluble.

C. *Metabolism.*  The metabolism of pyrrole and pyrrolidine, as well as of pyrrole- and pyrrolidine-containing molecules, follows the pattern expected for an aromatic compound and a secondary amine, respectively. Aromatic hydroxylation would be predicted for pyrrole, while pyrrolidine could be expected to undergo conjugation with glucuronic acid or sulfuric acid. Acetylation, a common reaction for secondary amines, might also be expected to occur.

### 3. Sulfur

*A. Nomenclature.* As with the previous oxygen and nitrogen heterocycles, two sulfur-containing heterocycles, THIOPHENE and TETRAHYDROTHIOPHENE, exist. Here again, the common nomenclature is used in most instances. With substituted analogs, the numbering of the rings starts with the heteroatom, proceeding either clockwise or counterclockwise, such that any substituent receives the next lowest number.

|              | Thiole | Thiolane |
|--------------|--------|----------|
| IUPAC :      | Thiole | Thiolane |
| Common:      | Thiophene | Tetrahydrothiophene |

*B. Physical-Chemical Properties.* The properties of the five-membered sulfur-containing heterocycles are based upon the proper recognition of the class of compounds to which they belong. Thiophene is an aromatic ring and is therefore relatively stable, while tetrahydrothiophene is a thioether. Unlike oxygen ethers, the thioethers are fairly stable compounds, and, also unlike the oxygen analogs, the sulfur-containing compounds are less water soluble. In general, the substitution of sulfur for oxygen results in a significant decrease in hydrophilic character and a corresponding increase in lipophilic character.

*C. Metabolism.* The predicted metabolic pattern for thiophene is aromatic hydroxylation, while for reduced thiophenes, oxidation of sulfur should occur. This was previously discussed for thioethers and is shown in Figure 15–6.

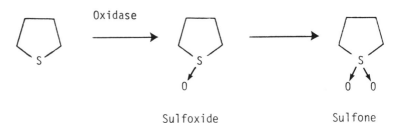

Sulfoxide                                    Sulfone

Fig. 15–6.   Metabolic oxidation of tetrahydrothiophene

# FIVE-MEMBERED RING HETEROCYCLES WITH TWO OR MORE HETEROATOMS

## 1. Oxygen and Nitrogen

*A. Nomenclature.* Two oxygen-plus-nitrogen heterocycles found in medicinal agents are OXAZOLE and ISOXAZOLE. The oxazole nomenclature is similar to its official name, but for convenience, the -1,3- is dropped, and it is understood that oxazole consists

| | | |
|---|---|---|
| IUPAC : | 1,3-Oxazole | 1,2-Oxazole |
| Common: | Oxazole | Isoxazole |

of the 1,3 arrangement of oxygen and nitrogen. The only other arrangement of these two atoms in a five-membered ring would be the 1, 2 placement. Since this is an isomer of oxazole, the common name is isoxazole. As shown previously, the numbering of the ring proceeds from oxygen at the 1 position to nitrogen such that the nitrogen is the next lowest position (i.e., 3 in oxazole or 2 in isoxazole).

The partially and totally reduced derivatives of oxazole that serve as basic nuclei in medicinally active chemicals are shown. The nomenclature is based upon the general rules presented in Table 15–2.

|  |  |
|---|---|
| IUPAC :   2-Oxazoline | Oxazolidine |

*B. Physical-Chemical Properties.* Both oxazole and isoxazole are aromatic compounds. The aromatic $\pi$ cloud is made up of two electrons from each double bond, plus a pair of electrons contributed by the oxygen atom. Since nitrogen is left with its unshared pair of electrons, both of these compounds are basic, although they should

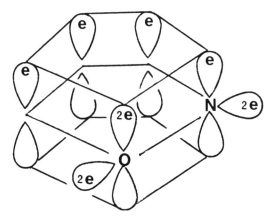

Fig. 15–7.  π electron structure of isoxazole

be recognized as weak bases (Fig. 15–7). Both compounds can be converted to salts with a strong acid such as hydrochloric acid. As their hydrochloric salts, oxazoles and isoxazoles would be predicted to be water soluble.

2-Oxazoline and oxazolidine are also basic compounds, but prove unstable in aqueous acid media. An example of such instability is shown in Figure 15–8.

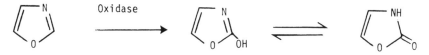

Fig. 15–8.  Acid-catalyzed hydrolysis of 2-oxazolidine

*C. Metabolism.*  The only characteristic metabolism of significance is aromatic hydroxylation. In the case of oxazole, the product formed exists in the "keto" form shown in Figure 15–9.

Fig. 15–9.  Metabolic hydroxylation of oxazole

### 2. Nitrogen and Nitrogen

*A. Nomenclature.*  Two important dinitrogen heterocycles are found in medicinal agents; these are PYRAZOLE and IMIDAZOLE. As with several previous examples, the student should be familiar with the common name, since it is used in most cases. The important partially reduced and saturated analogs are those of imidazole and

| IUPAC : | 1,2-Diazole | 1,3-Diazole |
|---|---|---|
| Common: | Pyrazole | Imidazole |

are 2-IMIDAZOLINE and IMIDAZOLIDINE, respectively. These compounds are numbered similarly to that shown for imidazole.

B. *Physical-Chemical Properties.* Pyrazole and imidazole are both aromatic compounds that have one basic nitrogen and a neutral nitrogen. The aromatic nature arises from the four $\pi$ electrons and the unshared pair of electrons on the -NH- nitrogen. Some care

Common:      2-Imidazoline            Imidazolidine

should be taken in specifying which nitrogen is basic since, in unsymmetrical derivatives, resonance prevents one from isolating a specific compound, as shown in Figure 15–10. All that can be said is that one nitrogen is basic.

In 2-imidazoline, both nitrogens are basic. The 2° amine is more basic than the sp² nitrogen. But again, it may not be possible to specify which nitrogen is sp³ and which is sp². Imidazolidine, on the other hand, is made up of two 2° amines and is quite basic. Im-

Fig. 15–10.  The imidazole heterocycle has one basic nitrogen, which, because of resonance, may be either nitrogen

idazolidine, like oxazolidine, is unstable in aqueous acid and is hydrolyzed to ethylenediamine and formaldehyde.

C. *Metabolism.* The predicted metabolism is uneventful. While the aromatic compounds may be prone to hydroxylation, the reduced heterocycles would be expected to act like secondary amines in vivo and undergo conjugation reactions.

### 3. Nitrogen and Sulfur

A. *Nomenclature.* The only heterocycle to be considered that contains a nitrogen and sulfur in a five-membered ring is the chemical 1,3-THIAZOLE. The nomenclature used is that derived from Tables 15−1 and 15−2.

1,3-Thiazole

B. *Physical-Chemical Properties.* The properties of thiazole are similar to those of oxazole. This compound is aromatic, and the nitrogen in this compound with its unshared pair of electrons is basic.

C. *Metabolism.* The metabolic properties are analogous to those of oxazole. Aromatic hydroxylation would be predicted for thiazole.

### 4. Complex Five-Membered Heterocycles

A. *Nomenclature.* A series of miscellaneous heterocycles important to medicinal chemistry are shown here. The TRIAZOLE nomenclature is self explanatory. Tri- means three, az- signifies nitrogen, and -ole means an unsaturated five-membered ring. The small "s" represents the symmetrical arrangement of nitrogens.

s-Triazole                1,3,4-Thiadiazole

The 1,3,4-THIADIAZOLE nomenclature is also self explanatory and follows the priority rules previously mentioned and the abbreviations for the heteroatoms (a sulfur, two nitrogens in a five-membered ring).

The OXAZOLIDINONE and OXAZOLIDINDIONE should be understandable based on oxazolidine nomenclature explained earlier, with the "one" signifying a carbonyl and "dione" representing two carbonyls at the 2 and 4 positions.

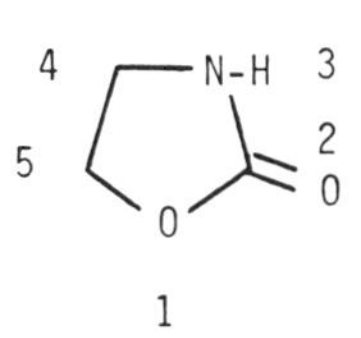

Oxazolidin-2-one 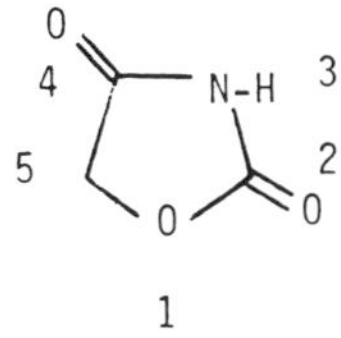     Oxazolidin-2,4-dione 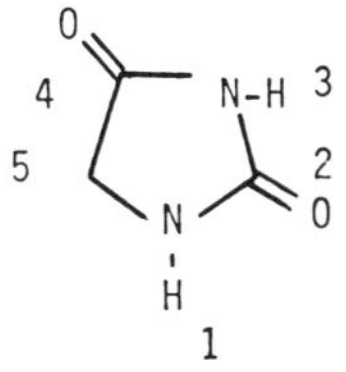     Hydantoin

The final important nucleus is the HYDANTOIN nucleus. This common nomenclature, a replacement for imidazolidindione, is used for an important class of anticonvulsants.

B. *Physical-Chemical Properties.* The triazole and thiadiazole are typical aromatic nuclei that have two basic nitrogens in the ring. Little additional information is necessary, since neither compound has any unique property that we need to be concerned with.

The oxazolidin-2-ones are cyclic analogs of a class of compounds discussed previously, namely, the carbamates. Like their straight-chain relatives, the oxazolidin-2-ones are readily hydrolyzed by acid or basic media (Fig. 15−11).

The oxazolidin-2,4-diones do have a chemical property unique to the "imide" portion of the structure (Fig. 15−12). Although an amide is neutral, the addition of a second carbonyl covalently bonded to the nitrogen produces the imide functional group, which has acidic properties. Two electron-withdrawing groups on either side of the -NH- group withdraw the unshared electron pair as well as the electron pair making up the nitrogen-hydrogen bond. This allows the hydrogen to be abstracted by a strong base, forming an alkaline salt that is quite water soluble.

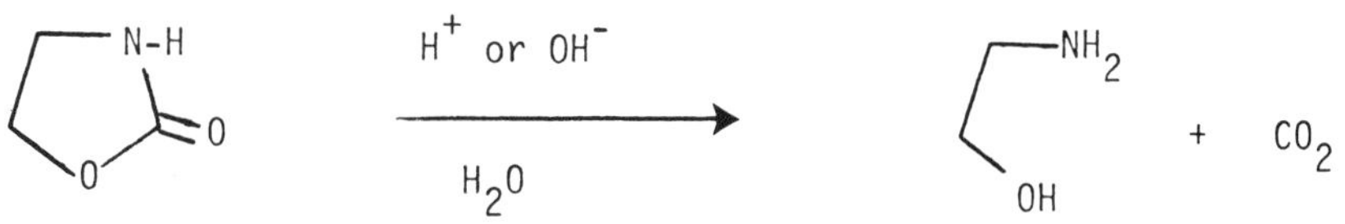

Fig. 15−11.  Acid- or base-catalyzed hydrolysis of oxazolidin-2-one

Fig. 15–12.  Salt formation at the imide nitrogen of oxazolidin-2,4-dione

The final heterocycle, the hydantoin, is a cyclic urea, but in addition it contains an imide functional group. The presence of the hydrogen on the imide nitrogen again results in a compound with acidic properties.

## SIX-MEMBERED RING HETEROCYCLES

### 1. Nitrogen

*A. Nomenclature.*   The important six-membered ring nitrogen containing heterocycles are PYRIDINE, the aromatic compound, and PIPERIDINE, the saturated compound.

|  | |
|---|---|
| IUPAC :  Azine | Hexahydroazine |
| Common:  Pyridine | Piperidine |

These two heterocycles are commonly found in medicinal agents, and in most cases the common name is used exclusively.

*B. Physical-Chemical Properties.*   Pyridine, unlike its carbon analog benzene, is quite water soluble. The explanation for this fact lies in the availability of an unshared pair of electrons found on the nitrogen. This polar compound can hydrogen bond to water through this pair of electrons. The availability of the electrons accounts for the other property of pyridine, which makes it different from pyrrole, namely, the basicity of pyridine. Pyridine has a $K_b$ of 2.3 × $10^{-9}$, which can be compared with the nearly neutral pyrrole's $K_b$ of 2.5 × $10^{-14}$; yet pyridine is much less basic than alkylamines, which have $K_b$s of approximately $10^{-4}$. Pyridine and substituted pyridines have approximately the same basicity as the aromatic amines such as aniline, $K_b = 0.42 \times 10^{-9}$. Thus, it would be difficult to predict a difference in basicity between the two nitrogens in the following compound:

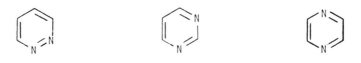

On the other hand, piperidine is nothing more than a cyclic alkyl amine. It is quite basic, with a $K_b$ of $2.0 \times 10^{-3}$. Other than the reactivity of pyridine and piperidine toward strong acids, one should consider both of these compounds as relatively stable.

C. *Metabolism.* Pyridine, since it is an aromatic compound, acts like the typical aromatic rings and undergoes hydroxylation. Piperidine acts like a typical secondary amine and would be predicted to undergo conjugation with glucuronic acid or sulfuric acid.

Dealkylation resulting in ring cleavage would not be expected to occur with piperidine, since dealkylation occurs primarily with amines that are substituted with smaller alkyl groups such as methyl or ethyl groups.

## SIX-MEMBERED RING HETEROCYCLES WITH TWO HETEROATOMS

### 1. Nitrogen and Nitrogen

A. *Nomenclature.* Several important heterocycles are formed from two nitrogens in a six-membered ring, and these are shown here with their respective nomenclature. Once again, the official nomenclature is usually neglected in favor of common nomenclature.

| IUPAC . | 1,2-Diazine | 1,3-Diazine | 1,4-Diazine |
|---|---|---|---|
| Common: | Pyridazine | Pyrimidine | Pyrazine |

B. *Physical-Chemical Properties.* All three of the compounds just described have properties similar to each other and similar to the properties of pyridine. The compounds are basic and of the same order of basicity as pyridine. Since these compounds are aromatic, they are also expected to be relatively nonreactive. Finally, parallel to the solubility properties of pyridine, these compounds are also water soluble.

### 2. Pyrimidines—I

A. *Nomenclature.* Three pyrimidine derivatives that are important to the structure of DNA and RNA, as well as to the structure of medicinally active agents, are shown here. While official nomenclature can be derived for these compounds, it is replaced with the

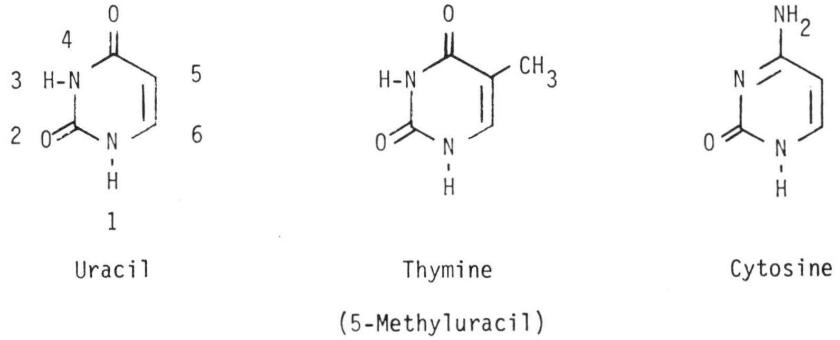

| Uracil | Thymine | Cytosine |
|--------|---------|----------|
|        | (5-Methyluracil) |   |

common names presented. Thymine may also be referred to as 5-methyluracil. Recognize the numbering system of these heterocycles. A clockwise direction is chosen such that the heteroatoms appear at the 1 and 3 positions and the carbonyls or carbonyl and amine are at the 2 and 4 positions. If the numbering were counterclockwise, the substituents on the ring would be at the 2 and 6 positions.

B. *Physical-Chemical Properties.* The substituted pyrimidines are complex molecules because of the nature of the substituents. Uracil and thymine may be considered to contain the neutral urea unit or the acidic imide moiety, as shown in Figure 15–13; but they can

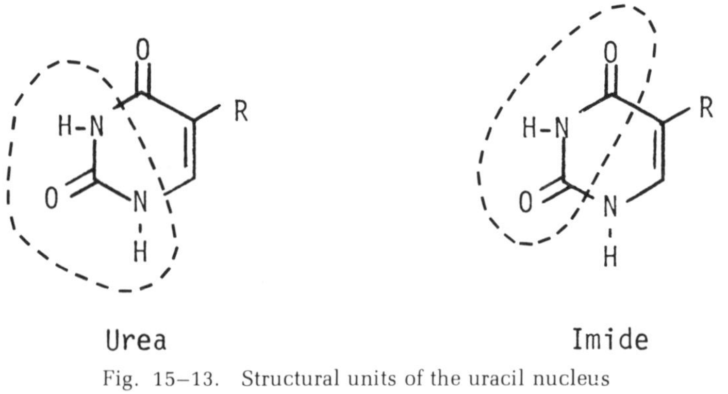

Urea                              Imide

Fig. 15–13.  Structural units of the uracil nucleus

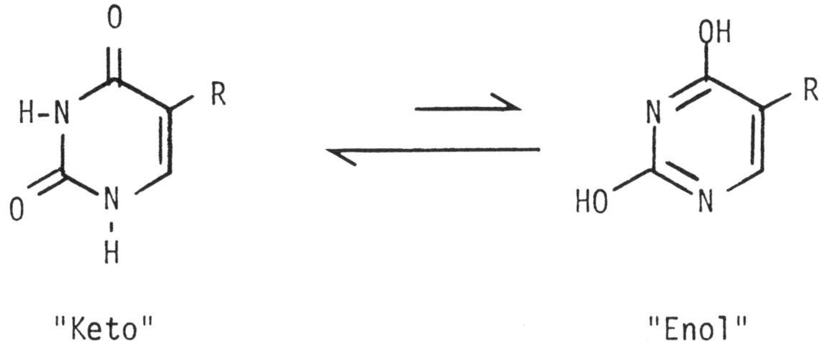
"Keto"
"Enol"

Fig. 15–14. "Keto"-"enol" equilibrium of the uracil ring

also be considered to exist in either the "keto" form or "enol" form, as shown in Figure 15–14. The "enol" form would be expected to have the acidic properties of a diphenolic compound and the basic properties of a pyrimidine. Since the compounds prefer the "keto" form, they are usually thought of as weak acids, but the weak acid-weak base properties of the "enol" form may account for the reduced solubility in water of uracil and thymine. Cytosine, with the 4-amino substituent and without an imide moiety, might be expected to be a weak base.

None of the substituted pyrimidines has any noteworthy instabilities.

*C. Metabolism.* The metabolism of these unique pyrimidines is important from the standpoint of both biochemical utilization of these compounds and drug metabolism of pyrimidine derivatives. Figure 15–15 outlines a pathway that converts uracil to a useful compound, uridylic acid, needed for the synthesis of RNA. In a similar manner to that shown for uracil, cytosine is conjugated with

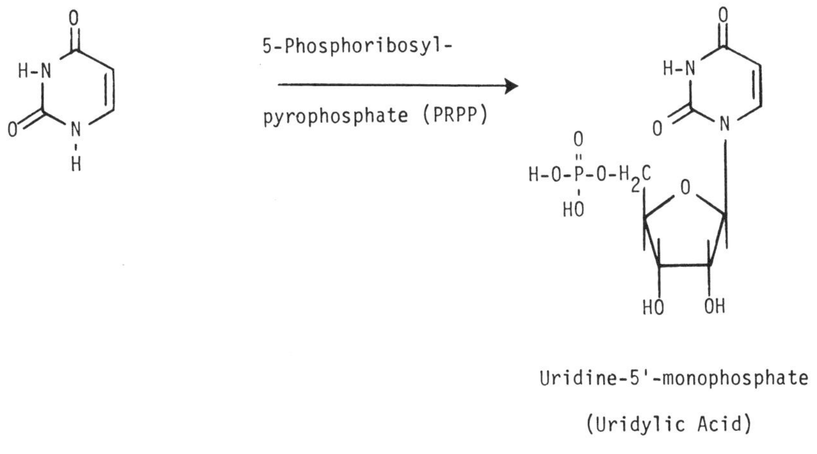
5-Phosphoribosyl-
pyrophosphate (PRPP)
Uridine-5'-monophosphate
(Uridylic Acid)

Fig. 15–15. Metabolism of uracil

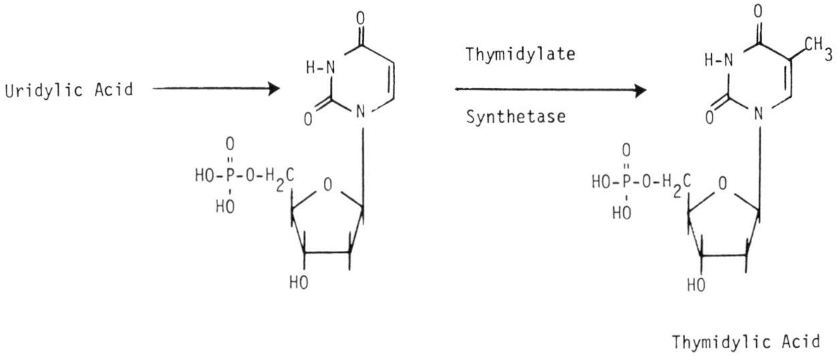

Fig. 15–16.  Biosynthesis of deoxythymidylic acid

PRPP to yield cytidine-5′-monophosphate (CMP) or cytidylic acid. Thymine is metabolized by conjugation, via a salvage pathway, with PRPP to the thymine ribosyl-5′-phosphate. This form of thymidylic acid can be utilized in specific RNA molecules. The biochemically important thymine deoxyribosyl-5′-phosphate is important in the biosynthesis of DNA but is derived from uridylic acid, which is first converted to deoxyuridylic acid and then into the deoxythymidylic acid (Fig. 15–16).

An example of a pyrimidine-substituted drug and its metabolic pattern is shown in Figure 15–17. 5-Fluorouracil can be conjugated with either ribose or deoxyribose, and the sugar is then phosphorylated to either 5-fluorouridine monophosphate (5-FUMP) or 5-fluorouridinedeoxyribosyl phosphate (5-FUDR). In this particular example, the 5-FUMP is responsible for the side effects of the drug, while 5-FUDR is responsible for the chemotherapeutic action of 5-FU.

### 3. Pyrimidines—II

*A. Nomenclature.*  Two special pyrimidines are barbituric acid and the substituted barbiturates. The substituted barbiturates represent a special class of compounds, which have been used for their

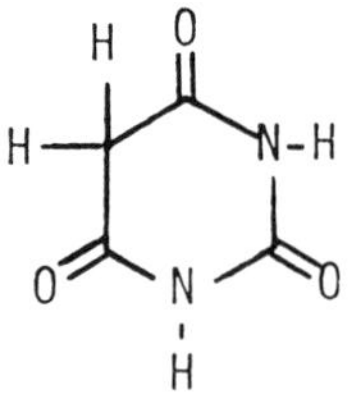

Barbituric Acid

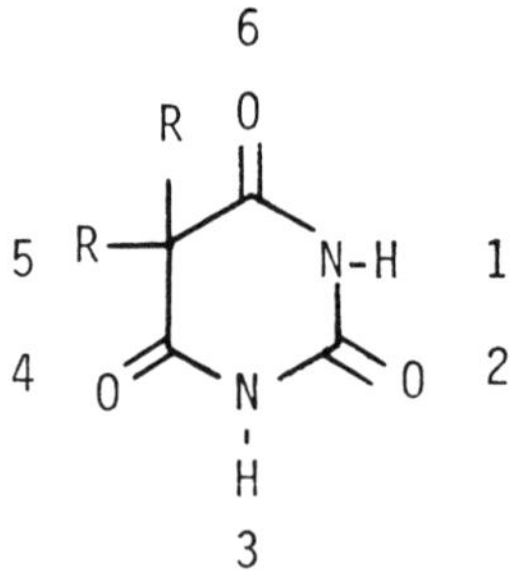

Barbiturates

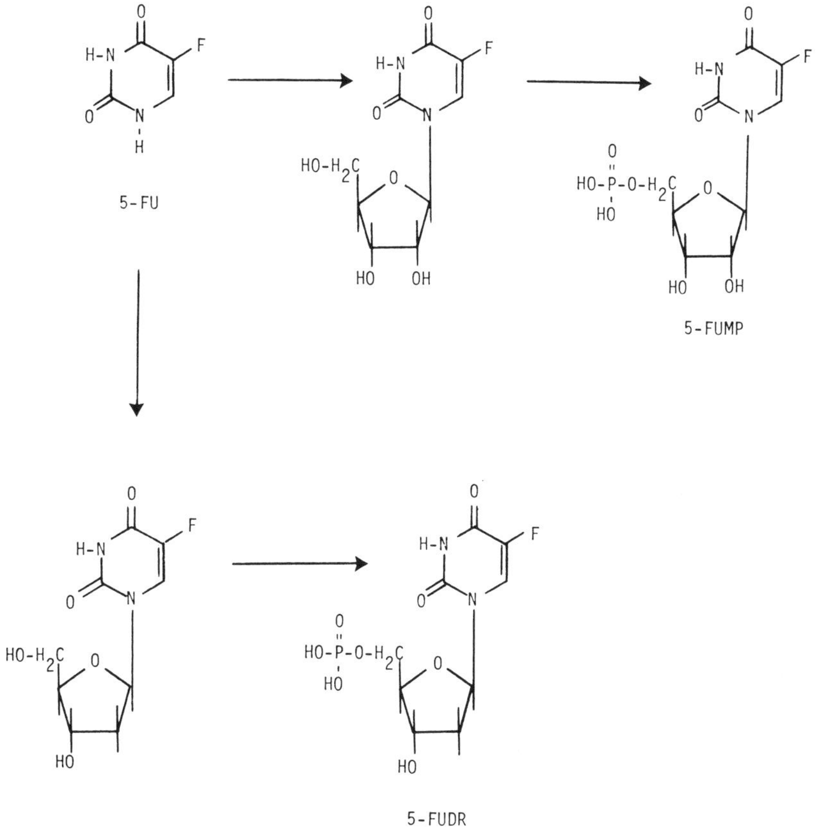

Fig. 15–17.   Metabolism of 5-fluorouracil

sedative-hypnotic action since the early 1900s. The numbering system starts with either nitrogen and proceeds such that the substituents (R) appear at the 5 position.

B. *Physical-Chemical Properties.*   Barbituric acid can exist in any of the four forms shown in Figure 15–18. Roentgenographic studies have shown that, as a solid, the compound exists in the trioxo form, while in solution, evidence rules against the trihydroxy form, but supports the other enolic forms as being present.

Barbituric acid is a fairly strong acid with a pK of 4.12, but upon substitution at the 5 position, the pK rises dramatically. The 5,5-disubstituted barbiturates have pK values of 7.1 to 8.1. Such compounds exist predominantly in the trioxo tautomeric form (Fig. 15–19). The 5,5-disubstituted barbiturates react with sodium hydroxide to form a salt that is quite water soluble (Fig. 15–20). Such salts when added to water will result in an aqueous medium that becomes quite alkaline owing to the fact that such a salt is made up of a weak

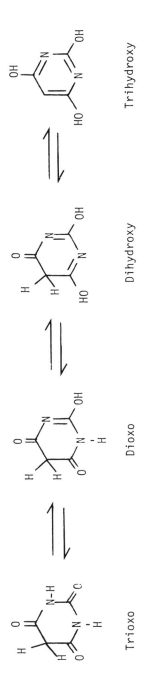

Fig. 15–18.   "Enol"-"keto" equilibrium of barbituric acid

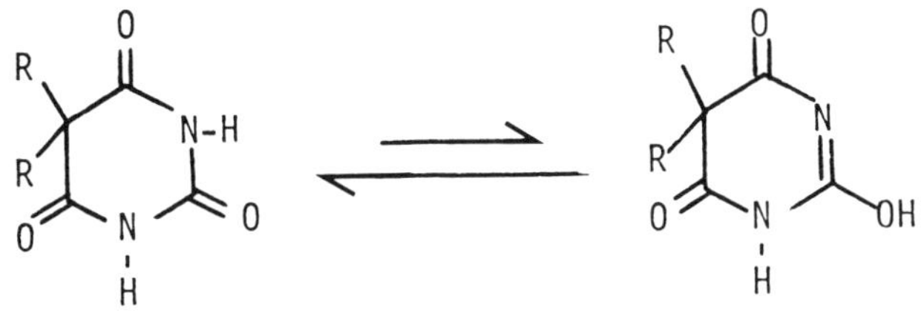

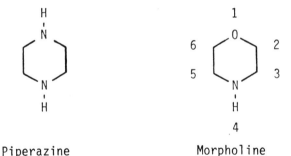

Fig. 15–19. "Enol"-"keto" equilibrium of 5,5-disubstituted barbituric acids

acid and a strong base. If the pH of the media is titrated to a neutral or acidic pH, the reaction will be reversed, resulting in precipitation of barbituric acid.

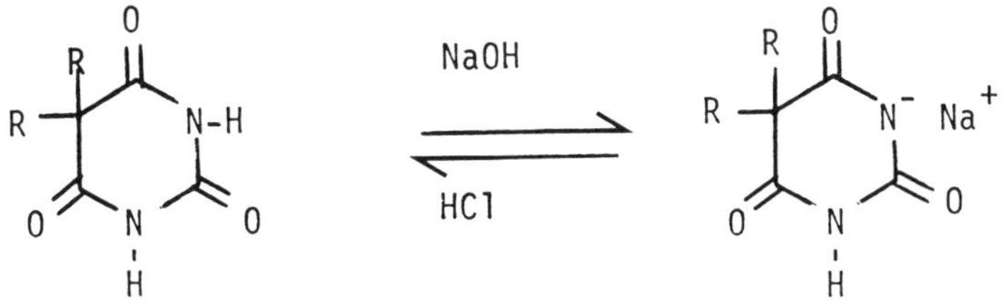

Fig. 15–20. Salt formation of 5,5-disubstituted barbituric acids

## 4. Saturated Six-Membered Heterocycles

*A. Nomenclature.* Two important saturated heterocycles that appear in drug molecules are PIPERAZINE and MORPHOLINE. The common names are used in most cases.

Piperazine                    Morpholine

*B. Physical-Chemical Properties.* Since these compounds are cyclic forms of a diamine, piperazine, and a secondary amine plus an ether, morpholine, the properties are the same as those reviewed in the chapters on these respective functional groups and need not be discussed at this point.

## SEVEN- AND EIGHT-MEMBERED RING HETEROCYCLES

### 1. Nitrogen

*A. Nomenclature.* Two heterocycles that complete our review of monocyclic heterocycles are HEXAHYDROAZEPINE and OCTA-

IUPAC :   Hexahydroazepine                 Octahydroazocine

HYDROAZOCINE. While the azepine has medicinal significance when it is part of a tricyclic ring system, the octahydroazocine is found in the drug guanethidine.

B. *Physical-Chemical Properties.* Both hexahydroazepine and octahydroazocine are cyclic secondary amines that are basic compounds and act like ordinary alkylamines.


## BICYCLIC HETEROCYCLES: FIVE-MEMBERED RING PLUS SIX-MEMBERED RING

### 1. One Nitrogen

A. *Nomenclature.* One important bicyclic ring system containing a single nitrogen is INDOLE. This nucleus is present in the amino acid tryptophan and is found in many alkaloids. Less important is the isomer of indole, ISOINDOLE, which is found in a partially reduced form in several drugs.

IUPAC :     Benzo [ b ] pyrrole              Benzo [ c ] pyrrole

Common:            Indole                      Isoindole

B. *Physical-Chemical Properties.* Indole is an aromatic compound with delocalization of the electrons across both rings (Fig. 15–21). Thus, like pyrrole, the benzopyrroles require the unshared pair of electrons on nitrogen to participate in the delocalized cloud of electrons. The result of this delocalization is that indole is a weak base and for our purposes will be considered neutral.

Indole and drugs containing the indole nucleus are easily oxidized when allowed to stand in contact with air. An indication of

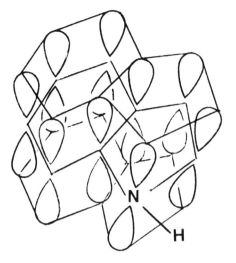

Fig. 15–21. π electron structure of indole

this reaction is the darkening of the color of the compound. It is best if indole-containing drugs are protected from atmospheric oxygen by storing under nitrogen.

C. *Metabolism.* Since indole is an aromatic nucleus, it is expected that aromatic hydroxylation would occur. Most indole-containing drugs are substituted at the 3 position, and the hydroxylation occurs at the 4-7 position of the molecule.

### 2. Two Heteroatoms

A. *Nomenclature.* Three bicyclic heterocycles that contain two heteroatoms are pictured here. The common nomenclature is based upon the name of the five-membered ring and since it is fused to a

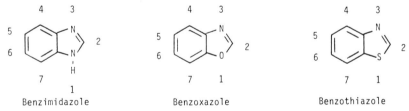

benzene ring, they are referred to as benz(o) ("o" is dropped when followed by a vowel) and then the name of the five-membered ring heterocycle. The numbering proceeds as shown. It should be noted that the bridgehead positions (e.g., the positions where the two rings join) are not numbered, because the carbons at the bridgehead are already fully substituted. In cases in which the benzene ring is reduced, the bridgehead position can be numbered at the 3a and 7a positions since they follow the 3 and 7 positions, respectively.

*B. Physical-Chemical Properties.* The properties of the benz-imidazole, benzoxazole, and benzothiazole do not differ significantly from the properties of imidazole, oxazole, or thiazole. All three compounds are aromatic, and all three have a weakly basic nitrogen in the molecule. The only property that does change is the fact that the molecules are less water soluble, since each has four additional carbon atoms present.

*C. Metabolism.* The predicted metabolism of these heterocycles is aromatic hydroxylation. The hydroxylation can occur at any of the positions occupied by hydrogen (2, 4, 5, 6, or 7 positions).

### 3. Four Heteroatoms

*A. Nomenclature.* An important bicyclic heterocycle is the PURINE nucleus. The purine can be thought of as a pyrimidine fused to an imidazole. The numbering follows this type of analogy. The six-membered ring is numbered first starting with one nitrogen atom and proceeding counterclockwise completely around the

| Adenine | Guanine | Xanthine |
|---|---|---|
| (6-Aminopurine) | (2-Amino-6-hydroxypurine) | (2,6-Purinedione) |

ring, including the bridgehead positions. This is then followed by numbering the five-membered ring.

Since three common substituted purines should be familiar to you, these compounds are also shown. They include the 6-aminopurine (adenine), 2-amino-6-hydroxypurine (guanine) (which actually exists not in the "enol" but rather in the "keto" form), and 2,6-purinedione (xanthine). All three of these compounds are common metabolites found in the human body.

*B. Physical-Chemical Properties.* Purine is an aromatic compound containing three basic nitrogens. Because of the ability of this

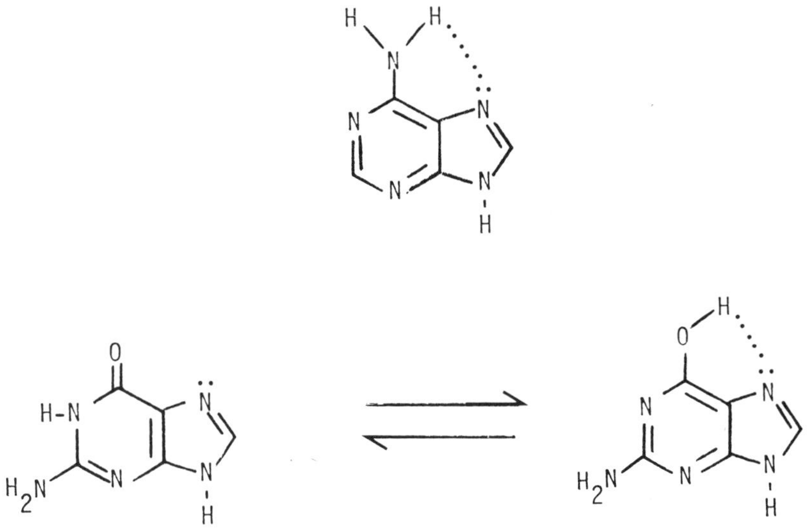

Fig. 15–22.   Intramolecular bonding present in adenine and guanine

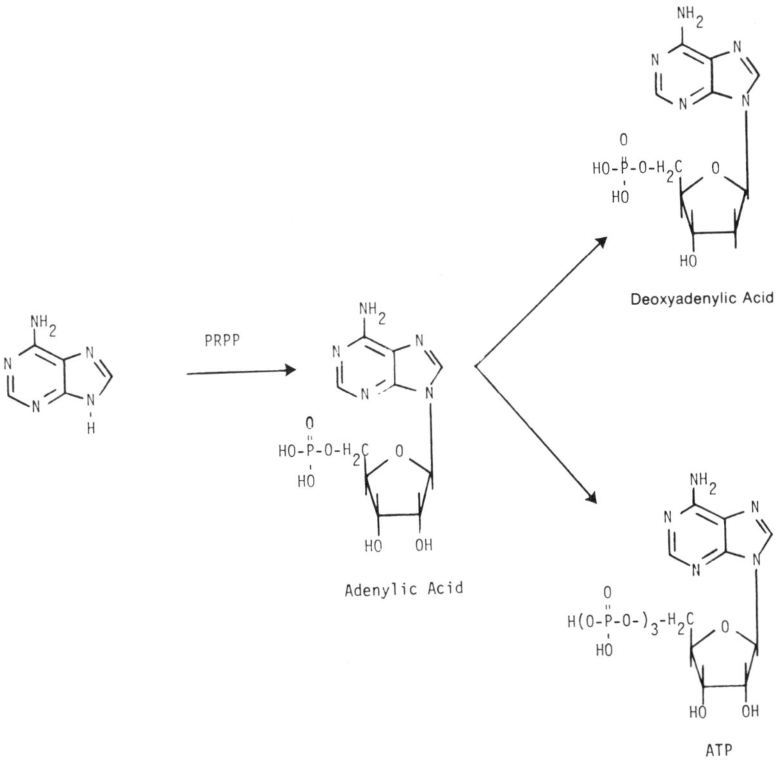

Fig. 15–23.   Metabolism of adenine

compound to bind to water through the unshared pair of electrons on the nitrogens, the compound is highly soluble in water. When the pyrimidine ring is substituted, as in the case of adenine, guanine, and xanthine, the water solubility decreases. This is probably due to intramolecular interactions, which will be discussed in the last chapter of this book. Let it suffice to say that intramolecular interactions such as those shown in Figure 15–22 decrease the attractions that can occur with water. Xanthine has basic properties due to one of the nitrogens in the imidazole ring and acidic properties due to the imide NH in the pyrimidine ring.

*C. Metabolism.* The metabolism of the substituted purine is quite systematic and is shown for adenine in Figure 15–23. The adenine is conjugated with 5-phosphoribosylpyrophosphate (PRPP) to give adenylic acid (adenosine-5'-phosphate). Adenylic acid in turn may be reduced to deoxyadenylic acid. A similar pattern of metabolism can lead to guanosine and xanthosine, which in turn can lead to guanylic acid and xanthylic acid.

A second type of metabolism common to purines is aromatic hydroxylation. An enzyme known as xanthine oxidase catalyzes this reaction. When xanthine is oxidized by xanthine oxidase, the resulting product is uric acid (Fig. 15–24).

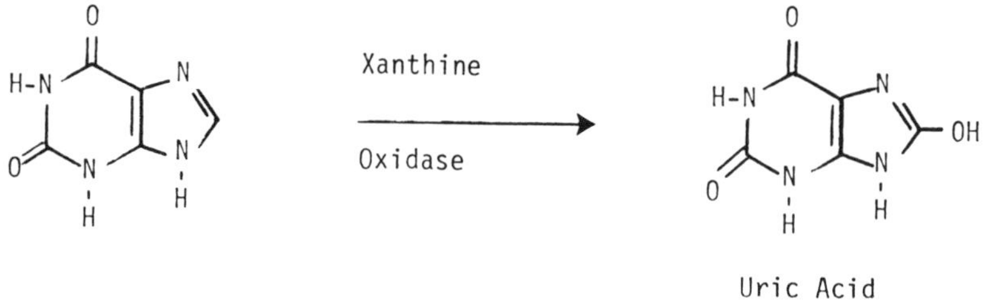

Fig. 15–24.   Metabolic oxidation of xanthine

## BICYCLIC HETEROCYCLES: SIX-MEMBERED RING PLUS SIX-MEMBERED RING

### 1. One Nitrogen

*A. Nomenclature.* Two important bicyclic heterocycles containing a nitrogen in two fused six-membered rings are QUINOLINE and ISOQUINOLINE. These nuclei are common to synthetic and naturally occurring drugs. The rings are numbered as shown. Note that the bridgehead positions are not numbered.

*B. Physical-Chemical Properties.* Both quinoline and isoquinoline are weak bases similar to pyridine. These weak bases will react with a strong acid such as sulfuric acid or hydrochloric acid to form water-soluble salts. Both compounds are aromatic and therefore have few additional properties that need concern us.

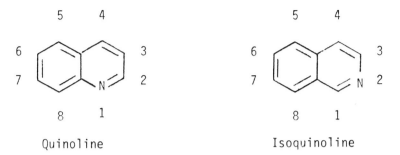

Quinoline                           Isoquinoline

*C. Metabolism.*   The common metabolism seen in quinolines and isoquinolines is aromatic hydroxylation at one of the positions occupied by hydrogen.

### 2. One Oxygen

*A. Nomenclature.*   Another important bicyclic heterocycle found in nature and several synthetic drugs is the COUMARIN molecule. Although the compound possesses a more complicated official name, the common name is usually used.

IUPAC :         2H-Benzopyran-2-one

Common:         Coumarin

*B. Physical-Chemical Properties.*   The coumarin molecule contains an intramolecular ester known as a lactone. Lactones experience the same types of instabilities as esters. Lactones are prone to hydrolysis catalyzed by either acid or base to give a carboxylic acid and phenol (Fig. 15–25).

$$H^+/H_2O$$

$$OH^-/H_2O$$

Fig. 15–25.   Acid- or base-catalyzed hydrolysis of coumarin

*C. Metabolism.*   Esterase-catalyzed hydrolysis of coumarins would be expected to occur in the body, the product being the more water soluble carboxylic acid shown in Figure 15–25.

### 3. Two or More Nitrogens

*A. Nomenclature.* Two additional bicyclic heterocycles that serve as nuclei for several synthetic drugs and natural products are QUINAZOLINE and PTERIDINE.

Quinazoline

Pteridine

*B. Physical-Chemical Properties.* The properties of quinazoline and pteridine are similar to the monocyclic six-membered heterocycles. Both compounds are aromatic and possess basic nitrogens. Like pyridine or pyrimidine, the nitrogens are weak bases and therefore require strong acids in order to form salts.

*C. Metabolism.* The metabolism expected for both quinazoline and pteridine is aromatic hydroxylation. This can occur at any of the positions occupied by a hydrogen.

## BICYCLIC HETEROCYCLES: SIX-MEMBERED RING PLUS SEVEN-MEMBERED RING

### 1. Two Nitrogen Atoms

*A. Nomenclature.* Two important six-plus-seven fused bicyclic heterocycles will be encountered in medicinal chemistry. Both of these systems are referred to generically as the BENZODIAZEPINES. The official nomenclature indicates that a benzene ring (benzo) has been fused to a seven-membered ring (pine), which in turn contains two nitrogens (diaz). The 1,4- designates the location of the two nitrogen atoms. Since a seven-membered ring can accommodate only three double bonds, the 3H tells indirectly that with a hydrogen at the 3 position, the double bonds are at the site of ring fusion as

3H-1,4-Benzodiazepine

1,3-Dihydro-2H-1,4-Benzodiazepin-2-one

well as at the 1,2 and 4,5 positions. An alternate arrangement of double bonds is shown for 1H-1,4-benzodiazepine. While 3H-1,4-benzodiazepine is the basic nucleus for the drug chlordiazepoxide, most of the benzodiazepines fall into the class of 1,3-dihydro-2H-1,4-benzodiazepin-2-ones. This heterocycle has an amide group present at the 1,2 position, with the carbonyl being represented by the -2-one nomenclature. The numbering system for the benzodiazepines is as shown.

1H-1,4-Benzodiazepine

*B. Physical-Chemical Properties.* There are few distinctive properties of the benzodiazepines that need concern us. The nitrogen at the 4 position is a basic, but only weakly basic, nitrogen. Salt formation at this position to give a water-soluble salt is usually not practiced, probably because of the weakness of this base.

*C. Metabolism.* Extensive metabolic data are available on the metabolism of the benzodiazepines. In many cases, the metabolism involves the additional substituents normally attached to the benzodiazepine nucleus. Metabolism of specific drugs will be discussed in the medicinal chemistry course. A common metabolic process that involves the 1,3-dihydro-2H-1,4-benzodiazepin-2-one nucleus is hydroxylation of the 3 position. This is seen with many of the anti-anxiety drugs.

## TRICYCLIC HETEROCYCLES

*A. Nomenclature.* Three tricyclic nuclei common to medicinal agents are PHENOTHIAZINE, DIBENZAZEPINE, and ACRIDINE. The nomenclature and numbering of these heterocycles are as shown. It should be noted that the numbering system for each of these compounds is different.

*B. Physical-Chemical Properties.* The phenothiazine nucleus contains a nitrogen that should be considered nearly neutral. Two aromatic rings attached to a nitrogen, each withdrawing electrons, reduce the basic property significantly. In most cases, this nitrogen will not form a salt with acid. The same reasoning holds for the nitrogen in 5H-dibenz[b,f]azepine. Acridine, although a weak base, can form salts with a strong acid.

An interesting physical property of the phenothiazine nucleus is that the molecule is not flat (Fig. 15–26). The shape of this molecule

Phenothiazine

5H-Dibenz [b,f] azepine

Acridine

is thought to affect its biologic activity, and the amount of bend from planarity therefore may be important.

A characteristic of the acridine nucleus is the fact that the molecule possesses color. The nature of the color will depend upon the substituents added to the three rings. The fact that a molecule possesses color indicates a highly conjugated molecule with alternating single and double bonds. With three conjugated rings, a yellow coloration is seen.

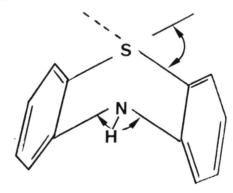

Fig. 15–26.  Conformational structure of phenothiazine

C. *Metabolism.*   The characteristic metabolism found in all three of the tricyclic compounds is aromatic hydroxylation. Since the medicinally useful agents have substitution on these nuclei, the substitution will influence the site of hydroxylation.

An additional metabolism common to the phenothiazine nucleus is oxidation of the sulfur to the sulfoxide or sulfone. This reaction can be expected for any thioether and is shown in Figure 14–1.

# 16

# Predicting Water Solubility

## 1. EMPIRIC METHOD

We have now reviewed the major functional groups that might be expected in drug molecules. It will soon become obvious to you that the majority of the drugs discussed are not simple monofunctional molecules but instead are polyfunctional molecules. Most drugs will be found to contain two, three, four, or more of the organic functional groups within a single chemical entity. How then does one predict physical and chemical properties of these more complex molecules? As mentioned throughout the book, one must recognize the individual functional groups within the more complex structures. Once this is done, the chemical properties, namely, in vitro stability and in vivo stability, are easily predicted. The chemical properties of a functional group are usually not affected by the presence of another functional group within the molecule. Therefore, each functional group can be treated independently of the other functional groups present.

If we consider the important physical property of water solubility, it is found that polyfunctional molecules behave somewhat differently than monofunctional molecules. A simple summation of the water-solubilizing properties of each functional group will usually not lead to a successful prediction of water solubility for the more complex systems. When one looks at the water-solubilizing property of a single functional group, there is no possibility of intramolecular bonding, that is, bonding within the molecule, because no second functional group is present. On the other hand, with polyfunctional molecules, intramolecular bonding may become a significant interaction. With the individual functional groups, the solubilizing potential of the groups took into consideration intermolecular bonding. As an example, an alcohol functional group in a molecule such as hexanol binds to a second molecule of hexanol through dipole-

dipole bonding. This bonding must be broken in order to dissolve the hexanol in water. When one states that an alcohol functional group solubilizes approximately six carbon atoms, this statement took into consideration intermolecular bonding of this type. But what about the polyfunctional molecules? The intermolecular bonding between like functional groups can still occur, but now a new type of bonding is possible, the intramolecular bond. Bonding may occur between dissimilar functional groups, and these types of intermolecular and intramolecular bonding may be quite strong. In order for a molecule to dissolve in water, the intramolecular and intermolecular bonding must first be broken so that the water molecules can bond to the functional groups.

Tyrosine

Solubility in water        0.45 g/1000 ml @ 25°C

An excellent example of the importance of intramolecular bonding is seen with the amino acid tyrosine. This molecule has three functional groups present, a phenol, an amine, and a carboxylic acid. By a simple summation of the water-solubilizing potential of each functional group, one would predict that the phenol would solubilize 6 to 7 carbon atoms, the amine 6 to 7 carbon atoms, and the carboxyl 5 to 6 carbon atoms, giving a total solubilizing potential of 17 to 20 carbon atoms. Tyrosine contains 9 carbons, yet the molecule is soluble to the extent of 0.5%. The explanation for this lack of water solubility can be understood if one recognizes the possibility of intramolecular bonding. The amino acid can exist as a zwitterion (Fig. 16–1). The charged molecule exhibits intramolecular ion-ion bonding. As a result, this destroys the ability of these two functional groups to bond to water. The phenol is not capable by itself of dissolving the molecule. If the intramolecular bonding is destroyed by either adding sodium hydroxide or hydrochloric acid to the amino acid, the resulting compound becomes quite water soluble.

Although less dramatic, most functional groups are capable of showing some intra- and intermolecular hydrogen bonding, which decreases the potential for promoting water solubility. How much

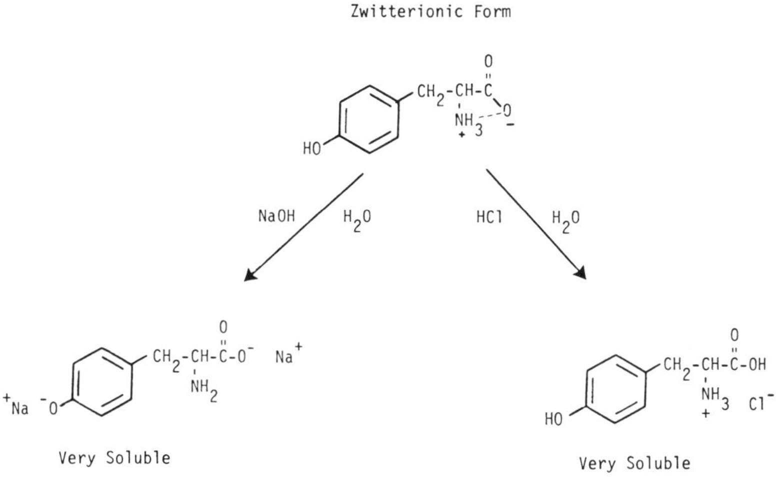

Zwitterionic Form

Fig. 16–1.   Solubilization of tyrosine in aqueous base or aqueous acid

weight should be given to each such interaction for individual functional groups? This is a difficult question to answer, but as a general rule, if one is conservative in the amount of solubilizing potential that is given to each functional group, one will find that fairly accurate predictions can be made for polyfunctional molecules.

In Table 16–1, the various functional groups that have been discussed are listed with the solubilizing potential of each group when present in a monofunctional molecule and the solubilizing potential when present in a polyfunctional molecule. This latter value will be the more useful value, since most of the molecules that we discuss will be polyfunctional.

Several examples will help demonstrate this method of predicting water solubility. In the first molecule (Fig. 16–2), one should recognize the presence of two tertiary amines. If the more liberal solubilizing potential for an amine is used, it might be expected that each amine would have the capability of solubilizing up to 7 carbon atoms, leading to a total potential of dissolving 14 carbon atoms in the molecule. Since the molecule contains 13 carbon atoms, one would predict that the molecule would be soluble. Using the more conservative estimate and allowing 3 carbons worth of solubility to each amine, a prediction of insoluble would result. It turns out that the molecule is water soluble. The use of the more liberal estimate in order to obtain the correct results is acceptable in this case since the molecule contains only amines that act alike, not creating any new inter- and intramolecular bonds.

With *para*-dimethylaminobenzaldehyde (Fig. 16–2), a nine-carbon molecule, the liberal estimate would predict solubility, since

Review of Organic Functional Groups

**Table 16–1.**
Water-Solubilizing Potential of Organic Functional Groups
When Present in a Mono- or Polyfunctional Molecule.
Water Solubility Is Defined As >1% Solubility.

| Functional Group | Monofunctional Molecule | Polyfunctional Molecule |
|---|---|---|
| Alcohol | 5 to 6 carbons | 3 to 4 carbons |
| Phenol | 6 to 7 carbons | 3 to 4 carbons |
| Ether | 4 to 5 carbons | 2 carbons |
| Aldehyde | 4 to 5 carbons | 2 carbons |
| Ketone | 5 to 6 carbons | 2 carbons |
| Amine | 6 to 7 carbons | 3 carbons |
| Carboxylic Acid | 5 to 6 carbons | 3 carbons |
| Ester | 6 carbons | 3 carbons |
| Amide | 6 carbons | 2 to 3 carbons |
| Urea, Carbonate, Carbamate | | 2 carbons |

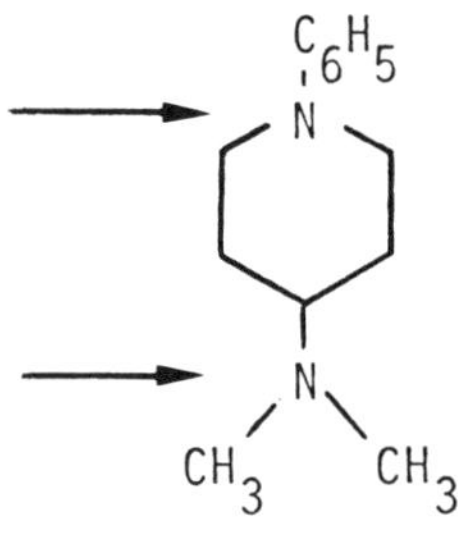

$C_{13}H_{20}N_2$

$7 + 7 = 14$

$3 + 3 = 6$

Water Soluble

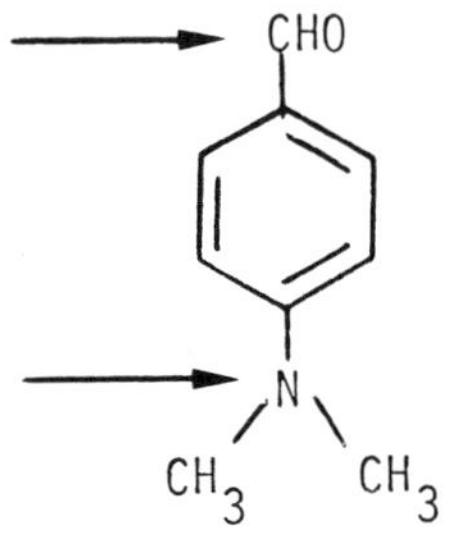

$C_9H_{11}NO$

$7 + 5 = 12$

$3 + 2 = 5$

Slightly Soluble

Fig. 16–2. Prediction of water solubility of organic molecules using mono- and polyfunctional estimates for the functional groups

the amine is capable of solubilizing up to seven carbon atoms and an aldehyde could solubilize up to five carbon atoms. On the other hand, the conservative estimate would predict insolubility with the amine worth three and the aldehyde worth two carbon atoms. This molecule is listed as slightly soluble, a result that falls between the two estimates. This simply shows that these are only predictions and, with borderline compounds, may lead to inaccurate results. The next examples (Fig. 16–3) lead to a more accurate prediction. In the first compound (Fig. 16–3), one should recognize the presence of

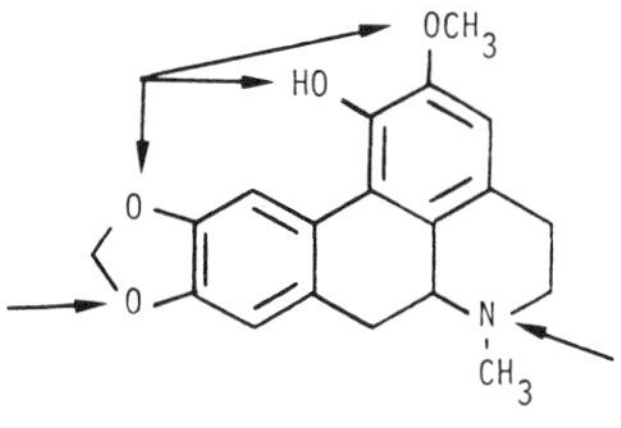

$C_{19}H_{19}NO_4$

$5 + 5 + 5 + 7 + 7 = 29$

$2 + 2 + 2 + 3 + 4 = 13$

Water Insoluble

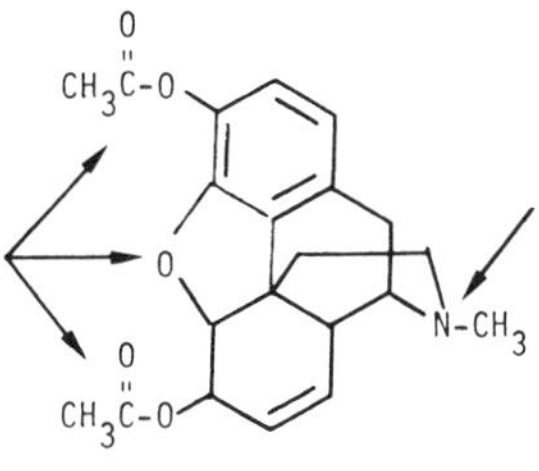

$C_{21}H_{23}NO_5$

$6 + 6 + 5 + 7 = 24$

$3 + 3 + 2 + 3 = 11$

Water Insoluble

Fig. 16–3. Prediction of water solubility of organic molecules using mono- and polyfunctional estimates for the functional groups

three ethers, a phenol, and a tertiary amine. Using the mono-functional solubilizing potential, one would expect enough solubility from these groups to dissolve this 19-carbon compound, since each ether would be assigned 5 carbons, the phenol 7 carbons, and the amine 7 carbons worth of solubilizing potential. If one uses the more conservative estimate, which takes into consideration the intra- and intermolecular bonding, however, each ether contributes two carbons worth of solubility, while the phenol and amine contribute three and four carbons worth of solubilizing potential, respectively. The prediction now is that the molecule is insoluble in water, and this turns out to be the case.

The next two examples use the same approach. The first compound (Fig. 16–4) has two ethers, two alcohols, and an ester. Using the liberal monofunctional estimates for water solubility would predict a soluble compound, while the conservative estimate would predict only 15 carbons worth of solubility. Since the compound possesses 15 carbons, one would predict solubility by either approach and the compound is soluble. In the last compound we should rec-

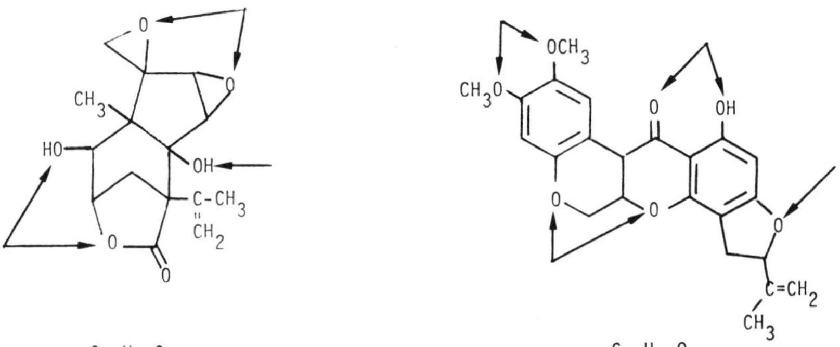

$C_{15}H_{18}O_6$

5 + 5 + 6 + 6 + 6 = 28

2 + 2 + 4 + 4 + 3 = 15

Water Soluble

$C_{23}H_{22}O_7$

5 X 5 = 25 + 6 + 7 = 38

5 X 2 = 10 + 2 + 6 = 18

Water Insoluble

Fig. 16–4. Prediction of water solubility of organic molecules using mono- and polyfunctional estimates for the functional groups

ognize the presence of five ethers, a ketone, and a phenol. The liberal estimate would result in a prediction of water solubility for this 23-carbon compound, but using the conservative estimate the more accurate prediction of water insolubility would result.

## 2. ANALYTIC METHOD

Throughout this presentation, emphasis has been placed on the water-solubilizing properties of the common organic functional groups. This was recapitulated in Table 16–1 with carbon-solubilizing potentials for each functional group, and the use of these values was demonstrated by the examples shown in Figures 16–2 through 16–4. While this approach is empiric, others have attempted to derive an analytic method for calculation of water solubility. One such mathematical approach recently reported by L. A. Cates (Am. J. Pharm. Ed., 45, 11, 1981) is now presented. This approach is based upon the partitioning of a drug between octanol (a

$$\log P = \frac{\text{Conc. of Drug in Octanol}}{\text{Conc. of Drug in Water}}$$

standard for lipophilic media) and water. The base-ten logarithm of the partition coefficients is defined as log P. While the measured log P values are a measure of the solubility characteristics of the whole molecule, one can use fragments of the whole molecule and assign a specific hydrophilic-lipophilic value (defined as $\pi$ value) to each of these fragments. Thus, a calculated log P can be obtained by the sum of the hydrophilic-lipophilic fragments:

$$\log P_{calc.} = \Sigma \pi \ (fragments)$$

To use this procedure, the student must fragment the molecule into basic units and assign an appropriate $\pi$ value corresponding to the atoms or groups of atoms present. Table 16–2 lists the common fragments found in organic molecules and their $\pi$ values. Positive values for $\pi$ mean that the fragment, relative to hydrogen, is lipophilic or favors solubility in octanol. A negative value indicates a hydrophilic group and thus an affinity for water. While the environment of the substituent can influence the $\pi$ value, such changes are small, and for our purposes this factor can be neglected.

Through the examination of a large number of experimentally obtained log P and solubility values, an arbitrary standard has been

**Table 16–2.**
Hydrophilic-lipophilic Values ($\pi$ Values) for
Organic Fragments

| Fragments | $\pi$ Value |
|---|---|
| C(aliphatic) | + 0.5 |
| Phenyl | + 2.0 |
| Cl | + 0.5 |
| $O_2NO$ | + 0.2 |
| IMHB | + 0.65 |
| S | 0.0 |
| O=C-O | – 0.7 |
| O=C-N (other than amine) | – 0.7 |
| O (hydroxyl, phenol, ether) | – 1.0 |
| N (amine) | – 1.0 |
| $O_2N$ (aliphatic) | – 0.85 |
| $O_2N$ (aromatic) | – 0.28 |

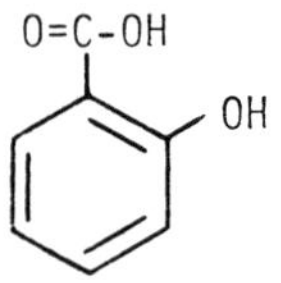

Solubility 0.2%

| Calc. log P without IMHB | | Calc. log P with IMHB | |
|---|---|---|---|
| Phenyl............... | + 2.0 | Phenyl............... | + 2.0 |
| OH.................... | - 1.0 | OH.................... | - 1.0 |
| O=C-0................ | - 0.7 | O=C-0................ | - 0.7 |
| | + 0.3 | IMHB................. | + 0.65 |
| | | | + 0.95 |

Prediction:    Soluble

Prediction:    Insoluble

Fig. 16–5.  Calculation of water solubility of salicylic acid without and with the intramolecular hydrogen bonding (IMHB) factor

adopted whereby those chemicals with a positive log P value over +0.5 are considered water insoluble (i.e., solubility is less than 3.3% in water—a definition for solubility used by the USP). Log P values less than +0.5 are considered to be water soluble.

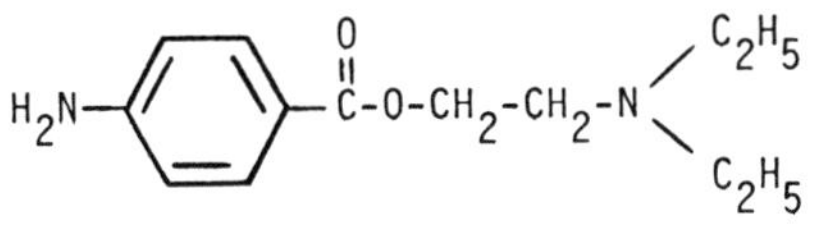

Procaine

| | |
|---|---|
| 6 - C @ + 0.5.............. | + 3.0 |
| Phenyl..................... | + 2.0 |
| 2 - N @ -1.0............... | - 2.0 |
| O=C- 0-.................... | - 0.7 |
| | + 2.3 |

Prediction:        Insoluble

Fig. 16–6.  Calculation of water solubility of procaine

This method of calculating water solubility has proved quite effective with a large number of organic molecules containing C, Cl, N, and O, but several additional factors may have to be considered for specific drugs. A complicating factor is the influence of intramolecular hydrogen bonding (IMHB) on $\pi$ values. As discussed in the previous empiric approach to predicting water solubility, IMHB would be expected to decrease water solubility, and, therefore, where IMHB exists, a $\pi$ value of +0.65 is added to the calculations. An example of using this factor is shown for salicylic acid (Fig. 16–5).

The log P values of a drug with acid or base character are influenced by the pH of the media in which the drug is placed. This is not surprising, since acid or base groups will become ionic under appropriate conditions. Although the $\pi$ values given in Table 16–2 were obtained under conditions in which the amine, phenol, or carboxylic acid are un-ionized, which would allow an accurate prediction of log P, observed log Ps at various pH values may not be accurate for water prediction. The experimental log Ps found for procaine are −0.32 (pH 7) and 0.14 (pH 8), both of which would lead to the prediction that procaine is water soluble. In fact, procaine is soluble to the extent of 0.5% at pH 7. The calculated log P = +2.3 (Fig. 16–6) correctly predicts that procaine is water insoluble.

# Appendix A

# Stereoisomerism—Asymmetric Molecules

A carbon atom with four different substituents does not possess a plane or point of symmetry and therefore is an asymmetric molecule. A carbon atom with two or more of the same substituents has either a plane or point of symmetry resulting in a symmetric molecule (Fig. A–1). For 2-methyl-2-butanol, carbon atoms 2,3 and 4 and the OH lie in a plane with the methyls and hydrogens symmetrically located before and behind the plane. This compound has a plane of symmetry and is therefore a symmetric molecule. On the other hand, 2-butanol does not have a plane of symmetry, is asymmetric, and consists of two molecules or a pair of enantiomers. The second enantiomer can be easily generated by reflecting the molecule in a mirror as shown in Figure A–2. If the mirror image is rotated by 180°, one can see that the enantiomers are not superimposable. 2-Butanol is said to be a *chiral* molecule with two enantiomeric forms. What is the significance of chirality? The two enantiomers have the same empirical formula and most physical-chemical properties are the same. The exception is that a chiral method of identification sees the two molecules as distinctly different. A common method of identifying enantiomers is by using plane polarized light. One of the isomers, when placed in a polarimeter, rotates the plane of polarization to the right (clockwise), is said to be dextrorotatory, and is labeled the *d* isomer or (+) isomer. The other isomer causes a counterclockwise rotation of the plane of polarization and is thus the levorotatory isomer abbreviated as the *l* isomer or (−) isomer. The degree of rotation is the same for both enantiomers but in opposite directions. The fact that enantiomers can bend plane polarized light has caused such compounds to be referred to as optically active

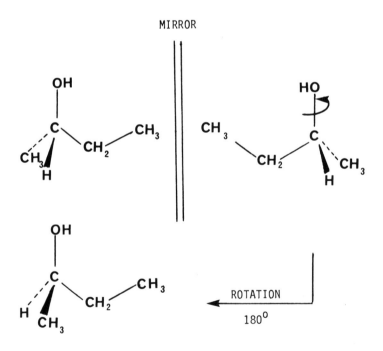

Fig. A–1.   Structure of the asymmetric 2-butanol and the symmetric 2-methyl-2-butanol

MIRROR

ROTATION

180°

Fig. A–2.   Enantiomers of 2-butanol

isomers. If a compound exists as an equal mixture of both isomers, the material is said to be racemic with a net rotation of polarization of zero. Other chiral substances, important to medicinal chemistry, that can often distinguish between enantiomers with profound differences, are biological enzymes. Since enzymes are proteinaceous in nature, they are made up of amino acids, which are chiral compounds. In the human body, enzymes are constructed of $\alpha$-amino acids. Many chiral enzymes react selectively with one of the enantiomers of a chiral drug, producing a biological response. The second enantiomer may have little or no biological activity. One must recognize the presence of a chiral center in a drug molecule and appreciate the importance of this property as it affects biological activity.

Finally, another aspect of a chiral center should be reviewed. The direction of rotation of plane polarized light is a relative property and does not indicate the absolute configuration around the chiral center. The Cahn-Ingold-Prelog "R and S" nomenclature is used to indicate absolute configuration. A set of arbitrary sequence rules assigns to the atoms around the chiral center priorities of 1 through 4, with number 1 being the highest priority. The molecule is then rotated so that the number 4 group is placed behind the remaining three groups and farthest from the eye. One then notes the direction in which the eye travels in going from 1 to 2 to 3. If the direction is clockwise, the molecule is assigned the absolute configuration of R, while if the direction is counterclockwise the center is assigned the S absolute configuration (Fig. A–3). While many sequence rules are

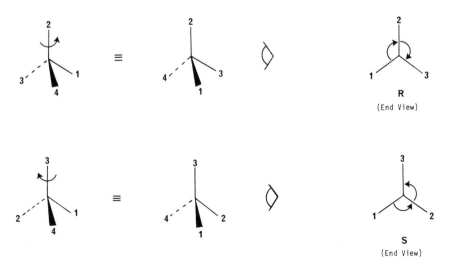

Fig. A–3.    Cahn-Ingold-Prelog method of assigning absolute configuration

$$N > C-O > C-H > H$$

Fig. A–4.   (S)-2-aminopropionic acid ((S)-alanine)

used for the many different functional groups encountered in organic chemistry, the one that will suffice for most situations is that
the atom with a higher atomic number precedes a lower atomic
number atom. Thus, for the amino acid alanine shown in Figure A–
4, the N (atomic No. 7) has higher priority than C (atomic No. 6),
which has higher priority than H (atomic No. 1). To differentiate
between the $CH_3$ and the COOH group, one must go to the atoms
attached to the carbons, and the O (atomic No. 8) has priority over H.

# Appendix B
# Acidity and Basicity

Throughout the book, considerable emphasis has been placed on the physical-chemical properties of the various functional groups. One of the major physical-chemical properties emphasized has been that of acidity/basicity. If a functional group is acidic, conversion of that group to a salt that can dissociate in water dramatically improves water solubility through ion-dipole bonding. In a similar fashion, if a functional group is basic, it can be converted to a salt by treatment with an acid. If the salt dissociates in water, water solubility will be increased through ion-dipole bonding. Since water solubility is quite important for drug delivery, it was felt that a short review of the concept of acidity and basicity was called for. In addition, a compilation of the important acids and bases, drawn from this book, will be presented in this appendix.

## DEFINITIONS OF ACIDS AND BASES

Although there are several definitions for acids and bases, the most useful for our purposes is the Brønsted-Lowry definition. According to this definition, an *acid* is defined as any substance that can *donate* a proton; a *base* is a substance that can *accept* a proton. Shown here is the reaction of HX with water. HX is donating a proton to water, and HX is therefore an acid. By virtue of the fact that water is accepting the proton, water is a base. The anion, $X^-$, formed from the acid, HX, is also capable of accepting a proton and is thus

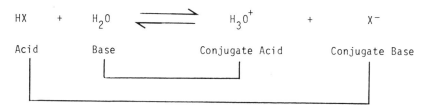

HX + $H_2O$ ⇌ $H_3O^+$ + $X^-$

Acid      Base      Conjugate Acid      Conjugate Base

defined as the *conjugate base* of HX. In a similar manner the hydronium ion, $H_3O^+$, is a *conjugate acid* of the base water. In this reaction, there are two conjugate acid-base pairs: the conjugate acid-base pair made up of HX and $X^-$; and the conjugate acid-base pair made up of $H_2O$ and $H_3O^+$. The Brønsted-Lowry definition of acids and bases necessitates the concept of conjugate acid-base pairs. Indeed, an acid (or base) cannot demonstrate its acidic (or basic) properties unless a base (or acid) is present. In the example shown,

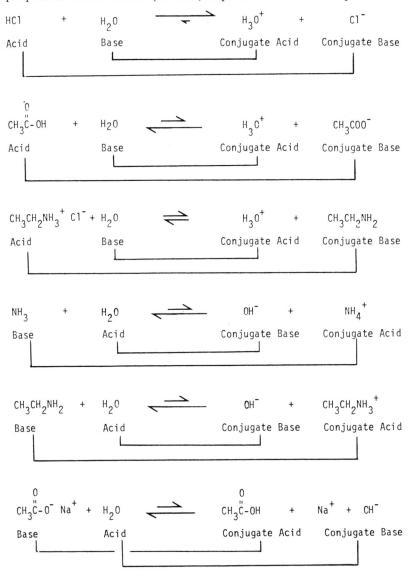

Fig. B–1.   Examples of conjugate acid-base pairs

HX cannot donate its proton unless there is another substance (a base) to accept that proton. Several additional examples of acids and bases are shown in Figure B–1.

Several interesting phenomena should be noted in these examples. Water is acting as a base in the first three examples and as an acid in the later three examples. Since water can act as either an acid or a base, it is said to be *amphoteric*. Also seen in Figure B–1 are examples of compounds that demonstrate another useful definition of a base. Lewis defined a base as an electron-pair donor. This definition is useful in identifying organic bases such as amines as possessing basic properties. Alkyl and aryl amines are basic by virtue of their ability to donate a pair of electrons.

## RELATIVE STRENGTHS OF ACIDS AND BASES

The strength of an acid depends on its ability to donate a proton. The strong acids have a strong tendency to give up a proton, while the weak acids have little tendency to give up a proton. Virtually all organic compounds could be considered acids. A compound like methane ($CH_4$) can give up a proton when treated with a sufficiently strong base. When dealing in the confines of drugs with the limitation that one is attempting to create salts with *water*-solubilizing potential, the list of acids is greatly reduced. Thus, the alcohol functional group, which an organic chemist may consider an acid, from our standpoint is considered a neutral functional group. The same limitations must be placed on the base. Many compounds that a chemist would consider basic will not give stable salts when placed in water. Therefore, the drug list of basic functional groups is quite limited.

Strong Acid :     HX     +     $H_2O$     $\rightleftharpoons$     $H_3O^+$     +     $X^-$

Weak Acid   :     HX     +     $H_2O$     $\rightleftharpoons$     $H_3O^+$     +     $X^-$

The discussion of the relative strength of an acid really becomes a discussion of the nature of the dissociation equilibrium for the acid in water. A strong acid is one that has a strong tendency to dissociate. The strong acids commonly used in pharmacy are nearly completely dissociated in water (e.g., $H_2SO_4$, HCl, $HNO_3$). If a com-

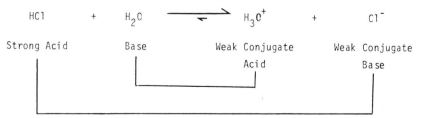

HCl     +     $H_2O$     $\rightleftharpoons$     $H_3O^+$     +     $Cl^-$

Strong Acid          Base          Weak Conjugate          Weak Conjugate
                                        Acid                  Base

pound is a strong acid, then its conjugate base is weak. The weak acids are those compounds that have a poor tendency to dissociate in water (e.g., carboxylic acids, phenols, sulfonamides, imides). Such compounds are characterized as having relatively strong conjugate bases. In all cases, the equilibrium will tend to favor the direction that gives the weaker acid and the weaker base.

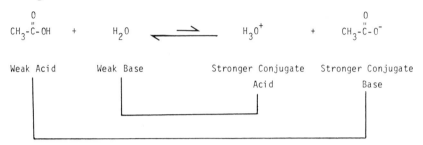

In a similar fashion, the relative strength of a base depends upon the ability of the chemical to give up a hydroxyl or the tendency to accept a proton. Table B–1 has a listing of common acids and bases in the order of acidity. The table actually becomes compressed when water is specified as our solvent. All of the acids above the hydronium ion show a reduction in apparent acidity because of the leveling effect in water. Since the strong acids (e.g., $H_2SO_4$, HCl, $HNO_3$) are nearly completely dissociated in water to form hydronium ions, their acidities become equal in water and the hydronium ion becomes the strongest ion. The leveling effect also affects strong bases. In water, the strongest base that can exist is the hydroxide ion; therefore, sodium hydroxide and potassium hydroxide, which completely dissociate, become equivalent in basicity.

We will comment briefly about water, which can dissociate to form hydronium ion and hydroxide ion, as shown. The hydronium ions and hydroxide ions formed by this dissociation are present in equal concentrations, i.e., there is no excess of either hydronium ions or hydroxide ions. Hence, water is neutral.

$$H_2O \quad + \quad H_2O \quad \rightleftharpoons \quad H_3O^+ \quad + \quad OH^-$$

| Acid | Base | Conjugate Acid | Conjugate Base |

### REACTION OF AN ACID WITH A BASE IN WATER

The reaction of an acid with a base in water is known as a neutralization reaction. When a strong acid reacts with a strong base, one

**Table B–1.**
Acid-base Chart

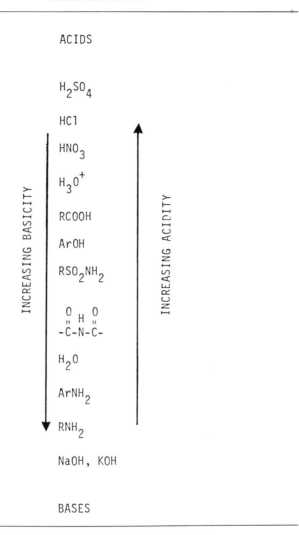

ACIDS

H$_2$SO$_4$

HCl

HNO$_3$

H$_3$O$^+$

RCOOH

ArOH

RSO$_2$NH$_2$

$$\overset{O}{\underset{\|}{-C}}-\overset{H}{N}-\overset{O}{\underset{\|}{C}}-$$

H$_2$O

ArNH$_2$

RNH$_2$

NaOH, KOH

BASES

INCREASING BASICITY

INCREASING ACIDITY

actually finds that the reaction consists of hydronium ions formed by the acid reacting with the hydroxide ions formed by the base. The neutralization reaction between a strong acid and a strong base results in an aqueous solution that is neutral. The anion of an acid and the cation of a base do not react with each other but are simply present as a salt. The situation is not the same when a weak acid is neutralized by a strong base. Since the weak acid exists primarily in the un-ionized form, the neutralization reaction is depicted for acetic acid and sodium hydroxide as shown below.

$$HNO_3 \quad + \quad H_2O \quad \rightleftharpoons \quad H_3O^+ \quad + \quad NO_3^-$$

$$KOH \quad + \quad H_2O \quad \rightleftharpoons \quad K^+ \quad + \quad OH^-$$

$$NO_3^+ \ + \ OH^- \ + \ K^+ \ + \ NO_3^- \ \longrightarrow \ H_2O \ + \ K^+ \ + \ NO_3^-$$

In the acid-base reaction between acetic acid and sodium hydroxide, sodium acetate, the weak conjugate base of acetic acid, and water, the weak conjugate acid of the hydroxide ion, are formed. The sodium acetate will partially dissociate in water, however, to afford acetic acid and hydroxide ion. This will result in a slight excess of hydroxide ion in solution when the neutralization reaction is complete, and the pH of the solution will be greater than 7, or alkaline.

$$CH_3COOH \quad + \quad NaOH \quad \rightleftharpoons \quad H_2O \quad + \quad CH_3COO^- \ Na^+$$

Weak Acid      Strong Base          Weak Acid      Weak Base

$$CH_3COO^- \ Na^+ \ + \ H_2O \quad \rightleftharpoons \quad CH_3COOH \quad + \quad OH^-$$

Base          Acid          Conjugate Acid      Conjugate Base

An analogous situation occurs in neutralization reactions between weak bases and strong acids, except that the final solution is acidic.

$$(CH_3CH_2)_3N \quad + \quad HCl \quad \longrightarrow \quad (CH_3CH_2)_3NH^+ \quad + \quad Cl^-$$

Weak Base      Strong Acid      Weak Acid      Weak Base

$$(CH_3CH_2)_3NH^+ \quad + \quad H_2O \quad \rightleftharpoons \quad (CH_3CH_2)_3N \quad + \quad H_3O^+$$

Acid          Base      Conjugate Base      Conjugate Acid

In this example, the dissociation of the salt formed during the neutralization reaction produces hydronium ions. The resulting solution from this neutralization reaction will have a pH less than 7 (acidic). The amount of hydronium ion formed depends on the extent of dissociation of the amine salt.

This concept is quite important if one considers what happens to the pH of water if a sodium or potassium salt of an organic acid is dissolved in water. The organic anion is the conjugate base of a weak acid and when placed in water the pH of the solution becomes alkaline. Another way of considering this is that, in water, one has the salt of a weak acid and a strong base. Since the strong base is farther

$$Na^+ \ ^-O-\overset{\overset{O}{\|}}{C}-R \quad + \quad H_2O \quad \rightleftharpoons \quad HO-\overset{\overset{O}{\|}}{C}-R \quad + \quad OH^- \quad + \quad Na^+$$

Weak Base                Acid                          Weak Acid            Strong Base

up the pH scale than the weak acid is down the scale, the net sum of this is that the pH remains greater than 7. When a sulfate, chloride, or nitrate salt of an organic base is dissolved in water, one has a salt of a weak base and a strong acid. The strong acid is farther down the pH scale than the base is up the scale, and the pH of the solution therefore is below 7. Using this concept, one can successfully predict the pH of many salts after dissolving the salt in water (Table B–2).

**Table B–2.**
Dissociation of Salts in Water

|  |  |  |  | Prediction |
|---|---|---|---|---|
| $ZnCl_2$ + $H_2O$ $\rightleftharpoons$ $H_3O^+$ + $Cl^-$ + $Zn(OH)Cl$ | | | | Acidic |
| Acid   Base   Strong Acid   Weak Base | | | | |
| $Na_2CO_3$ + $H_2O$ $\rightleftharpoons$ $Na^+$ + $OH^-$ + $H_2CO_3$ | | | | Basic |
| Base   Acid   Strong Base   Weak Acid | | | | |
| $Ca(NO_3)_2$ + $H_2O$ $\rightleftharpoons$ $H_3O^+$ + $NO_3^-$ + $Ca(OH)NO_3$ | | | | Acidic |
| Acid   Base   Strong Acid   Weak Base | | | | |

$$\text{(ring-}N^-Na^+\text{)} \quad + \quad H_2O \quad \rightleftharpoons \quad Na^+ \quad + \quad OH^- \quad + \quad \text{(ring-}N-H\text{)}$$

Base                     Acid              Strong Base       Weak Acid       Basic

The clue to predicting the correct answer is in being able to recognize whether one is dealing with salts of strong acids and weak bases or weak acids and strong bases. A rule of thumb that is most helpful is that the strong acids are hydrochloric, sulfuric, nitric, perchloric, and phosphoric acid. All other acids are weak. The strong bases are

$$R_3NH^+ \quad + \quad H_2O \quad \rightleftharpoons \quad R_3N \quad + \quad H_3O^+$$

Weak Acid        Base              Weak Base         Stronger Acid

sodium hydroxide and potassium hydroxide. All other bases are weak.

## ACIDIC AND BASIC ORGANIC FUNCTIONAL GROUPS

This section presents, in a single location, a synopsis of most of the organic acids and bases that are important in drugs as potential salt-forming sites. As stated earlier, nearly all organic compounds could be considered potentially acidic, but put in the context of water as the solvent, only a few functional groups are acidic enough to be of practical value.

Table B–3 lists the organic acids that have been reviewed in this book. Although sulfonic acids are the most acidic of the "organic" functional groups, few drugs contain them. The carboxylic acids are

**Table B–3.**
Order of Acidity of Organic Acids

| Acids | Acidity Constant in water $(K_a)$ |
|---|---|
| $R-SO_3H$ | $1 \times 10^{-1}$ |
| RCOOH | $1 \times 10^{-5} \longrightarrow 1 \times 10^{-4}$ |
| $Ar-\overset{O}{\underset{O}{\overset{\|}{\underset{\|}{S}}}}-NH-R$ | $1 \times 10^{-9} \longrightarrow 1 \times 10^{-6}$ |
| ArOH | $1 \times 10^{-11} \longrightarrow 1 \times 10^{-8}$ |
| (imide N-H structure) | $1 \times 10^{-9} \longrightarrow 1 \times 10^{-8}$ |

the most acidic organic functional groups found in drug molecules. The nature of the "R" group strongly influences the strength of this acidity (for a review of this influence, see Chapter 11). When "R" is an electron-withdrawing group, relative to hydrogen, the acidity increases, and when "R" is an electron-releasing group, relative to hydrogen, the acidity decreases. The effects of "R" on acidity, inductively or by resonance, stabilize or destabilize the carboxylate anion.

The sulfonamide is an acidic functional group provided that the sulfonamide is unsubstituted or monosubstituted. A disubstituted sulfonamide does not have a proton on the nitrogen and cannot be acidic. Phenols are relatively weak acids that can be strongly influenced by the nature of the aromatic substituent (see Chapter 7). Electron-withdrawing groups in the para or meta position increase acidity by resonance or a combination of resonance and inductive action. Electron-releasing groups likewise reduce acidity because such groups destabilize the phenolate anion. Finally, common to many heterocycles is the imide functional group, which must have a proton on the nitrogen to be acidic.

All of the organic acids (with the exception of sulfonic acid) are weak acids, which is to say that, in water, the equilibrium will favor the undissociated molecule. This also explains why salts of organic acids and strong bases, when dissolved in water, produce an appreciable quantity of hydroxide ion, and the aqueous solution is

$$Y-H \quad + \quad H_2O \quad \rightleftharpoons \quad H_3O^+ \quad + \quad Y^-$$

| Weak Acid | Weak Base | | Stronger Acid | Stronger Base |
|---|---|---|---|---|

alkaline. The degree of alkalinity will depend on the strength of the acid. Alkaline metal salts of phenols will result in a very basic aqueous solution, while alkaline metal salts of carboxylic acids are less basic.

Table B–4 lists the organic bases that have been reviewed in this book. The alkylamines are the most basic amines with the usual

**Table B–4.**
Order of Basicity of Organic Bases

| Bases | Basicity Constant in Water ($K_b$) |
|---|---|
| $R-NH_2, R_2NH, R_3N$ | $1 \times 10^{-4} \longrightarrow 1 \times 10^{-3}$ |
| $ArNH_2$ | $1 \times 10^{-13} \longrightarrow 1 \times 10^{-9}$ |
| N: | $1 \times 10^{-13} \longrightarrow 1 \times 10^{-10}$ |

order being secondary amines are more basic than tertiary amines which are more basic than primary amines. Aromatic amines are significantly $(10^{-10})$ less basic than alkylamines $(10^{-4})$, owing to the fact that electron-releasing groups (alkyls) increase basicity while electron-withdrawing groups (aryls) reduce basicity (Chapter 10). Aromatic heterocycles containing a nitrogen with a pair of nonbonding electrons are also basic; however, these compounds are weak bases.

$$B: \quad + \quad H_2O \quad \rightleftharpoons \quad BH^+ \quad + \quad OH^-$$

Weak Base      Weak Acid      Stronger Acid      Stronger Base

All of the organic bases are weak bases, which is to say that in water the equilibrium will favor the free base. Thus, when salts of the organic bases made with strong acids are dissolved in water, an appreciable quantity of hydronium ion will exist and the pH of the aqueous solution will be acidic, the degree of acidity depending on the strength of the base.

---

## ANSWERS TO REVIEW QUESTIONS

| | | | |
|---|---|---|---|
| 1. 3 | 11. 1 | 21. 4 | 32. 4 |
| 2. 4 | 12. 1 | 22. 1 | 33. 3 |
| 3. 2 | 13. 1 | 23. 2 | 34. 2 |
| | | 24. 2 & 5 | 35. 1 |
| 4. 2 | 14. 2 | | |
| 5. 2 & 4 | 15. 2 | 25. 5 | |
| 6. 3 | 16. 2 & 3 | 26. 1 & 3 | |
| | | 27. 1 | |
| 7. 2 & 5 | 17. 2, 3 & 4 | 28. 3 | |
| 8. 1 | 18. 4 | 29. 4 | |
| 9. 1 | 19. 3 | 30. 5 | |
| 10. 1 | 20. 5 | 31. 1 | |

---

# Index

Page numbers in *italics* refer to figures; page numbers followed by "t" indicate tabular material.

137